The Social Behaviour of Animals

THE SOCIAL BEHAVIOUR OF ANIMALS

STUART J. DIMOND
Lecturer in Psychology
University College of South Wales
and Monmouthshire, Cardiff

B. T. BATSFORD LTD London

First published 1970
© Stuart J. Dimond 1970
7134 0957 6

Made and printed in Great Britain by
William Clowes and Sons Ltd, London and Beccles
for the publishers B. T. Batsford Ltd
4 Fitzhardinge Street, London W.1

Contents

Acknowledgments

I am greatly indebted to all those friends and colleagues who through discussion and generous argument have contributed to the writing of this book. I would like to thank G. Westby particularly for his valuable argument and encouragement. I especially thank those friends who read parts of the manuscript, H. Smith who critically read the chapter on Social Learning, and A. Jolley who critically read the complete manuscript. W. Sluckin also read the book in full and it has benefited greatly from his wisdom. I am greatly obliged to him for his comments and for the way in which he has pointed to new perspectives in the problems which were encountered. Finally, I would like to thank my wife, who through her patience and invaluable help has made this work possible.

Introduction

The social behaviour of animals was one of the major concerns of the early naturalists. They were interested, not only in the behaviour of individuals in social contexts, but also in that shown when animals are aggregated together in large groups. The problems of the social behaviour of animals possess an intrinsic interest. This early work forms a considerable contribution to our knowledge of animal behaviour, but in recent years the methods have changed and the type of investigation undertaken has become considerably more sophisticated.

Experimental methods are of the very first importance. These bring to light the relationships between one aspect of behaviour and another, and they permit the investigation of causes. This book describes the development of recent methods—amongst them the ones used to study operant behaviour.

Skinner (1938) distinguishes between two types of behaviour:

1. Respondent behaviour, in which the organism reacts to a particular environmental stimulus;
2. Operant behaviour, in which the behaviour is emitted by the organism.

Emitted behaviour is brought under experimental control by judicious reinforcement. The experimenter observes the behaviour of the organisms and establishes control by reinforcing some patterns of behaviour but not others. Reinforcement traditionally takes the form of food reward, but many other types of reinforcement are possible.

An interest may, for example, be expressed in the way an animal learns to discriminate one shape from another. To study discrimination a pigeon could be required to peck at one key of a particular shape but to avoid pecking other keys. The pigeon is placed in a cage of restricted area. Each time it moves towards the appropriate key it is provided with food reward. The animal is reinforced for movement towards the key. More and more movement is required on each subsequent occasion until ultimately the pigeon approaches the correct key and pecks it to obtain food. The key itself is commonly used as the switch which operates the food magazine. It can also be used to trigger other complex electronic equipment. In this way the pigeon itself becomes an important link in the system. The machine itself is then used to study the pigeon's behaviour. The pigeon's pecking behaviour is automatically registered, recorded and analysed, and the subsequent events of the environment, e.g. the presence of food reward or particular patterns of stimulation, can be programmed to be dependent upon the nature of the animal's response. Rats, guinea-pigs, chimpanzees, human beings and other organisms can be used as links in automatic apparatus, which acts not only to record behaviour but also to establish control over the nature of their response.

This linking-in of an organism with a machine to promote training is most commonly thought of as a means of studying learning. It is in this area that this kind of method has had its greatest impact. Recently, however, the techniques and methods used in the study of operant conditioning have been much more widely applied, and it is their

use in the study of social behaviour which concerns us here.
A female monkey, for example, will press a lever opening
a window in order to view her own infant. A female laboratory rat in heat will press a lever in order to be allowed to
associate with the male. There are many examples of the
use of this kind of method in the study of animal behaviour. These will be discussed fully in later chapters.

The importance of operant behaviour in this respect lies
in the fact that the organism can be used as a link in an
electronic system, not to provide it with food reinforcement, but to examine the way in which it responds to a
wide variety of important social stimuli and conditions.

Apart from the advantage of automatic scoring and control, the method does have the additional advantage that
it allows an analysis of social behaviour to take place in
terms of an objective measure. The extent of lever pressing
provides an index of the value to the animal of the social
stimulus. In this way an analysis can be undertaken of those
social stimuli and situations which an animal will work to
engage or to perpetuate. Also, the character of the intrinsic
motivation governing its behaviour can be determined,
and the analysis of social behaviour by the use of operant
techniques allows us to extend considerably our knowledge
of this area.

Operant techniques have been used to throw new light
on the nature of social learning. This is an important area
of investigation. Social learning relates to two major areas
of research; first, the effects of early experience; and second,
the study of social motivation.

Early experience is important because the course of
early socialization may be radically changed through its
effects, and through its influence many patterns of social
behaviour appear to be laid down. The study of social
motivation is also important because it is intricately re-

lated to the problems of explanation and the reasons why animals behave as they do. Learning is an explanatory principle of the first order in relation to the problems of social behaviour and choice, as well as the problems of the acquisition of motor habits which are traditionally more closely associated with it. The study of motivation points to the importance of social rewards. Animals behave in a directed fashion to obtain these. They work in particular ways in order to view other animals and they may make strenuous efforts to rejoin companions from which they have been separated. This suggests something of the power which is implicit in the nature of social rewards.

It appears probable that we will be led to new ways of thinking about learning by the study of social behaviour.

It is already apparent that the traditional categories of reward which were supposed to promote learning in animals are too restrictive and narrowly based. These traditional reward systems are no longer sufficient, and recent discoveries in the field of social learning suggest that radical revision is necessary in our views about the processes which promote learning in animals and man.

It is not the intention to suggest, however, that all social behaviour is learned. The old debate as to what is learned and what is innate seems to be largely unproductive in this respect. The purpose is to emphasize the importance of social learning without denying the contribution made by other factors or determinants of behaviour.

Whilst it is the purpose of this book to describe the contemporary experimental approach to the study of animal behaviour and to deal with the nature of the investigations of operant behaviour, there are many other areas of enquiry in relation to the social behaviour of animals which cannot be ignored. One of these areas concerns the problems of communication and language.

It is clear that any attempt to describe animals as dumb brutes guided only by blind and irrational instincts, completely lacking the ability to relate the events of the environment to one another, and lacking in social feeling or awareness, is a picture of animal behaviour which cannot be credibly maintained. This is nowhere more apparent than in the study of the development of language and communication. Attention is drawn to the complexities of communication in the interaction between one animal and another, and the importance of the study of communication in understanding the social behaviour of animals is emphasized.

A large part of this book is concerned with the behaviour which individual animals show when they are placed within a social context. This behaviour can be readily studied by the use of experimental techniques, but we are also concerned with group process and here the methods of investigation and means of analysis become considerably more complicated. Not all the topics concerned with the social behaviour of animals can be treated in a purely experimental fashion, and here natural observation forms an important addition to the experimental approach. In these complex areas the problems often lend themselves more readily to the techniques of natural observation.

The areas of research which have been particularly closely associated with the use of natural observation are described in the chapters on Status Relationships and Group Behaviour. These deal with investigations of animal behaviour which might be described as sociological and which are concerned with the organization and structure of groups of animals. Experimental methods are not entirely ignored in these chapters, however, and it is argued that those experimental investigations which do exist foreshadow the development of an approach to group

behaviour through the study of animals which is directly
sociological in nature but which is also frankly experimental.

Finally, we have to consider the way in which studies
of animal behaviour relate to the behaviour of man. During the processes of physical growth and development there
are changes which come about in behaviour. It is possible
to study physical development from the fertilized egg stage
to that of the adult animal. Behaviour may be also studied
by looking at early forms, in order to see how these change
and mature into the ultimate adult product. In this way
it is possible to trace behavioural development through its
complete sequence, to see how adult behaviour is derived
from more primitive forms.

We have become accustomed to enquiries conducted
upon animals which have relevance to some aspect of
human function. Drugs which are to be administered to
human beings are commonly tested first on animals to determine the effects they may have upon living systems.
When human beings are required to penetrate into unusual or difficult environments, such as those concerned,
for example, with radioactive hazards or unusual gravita-
tional stress, it is common practice to investigate first the
effects of these environments upon animals. Dogs may be
used as a vanguard for exploration, as occurred in the development of the Russian space programme. If the environment proves itself to be safe for these living creatures, then
there is a reasonable chance that it will also prove suitable
for occupation by human beings, and so the risk to man is
minimized. We have long become used to the study of
animals in medical research where, for example, organ
transplants may be carried out to develop the appropriate
surgical and related techniques, so that organ transplant
may take place safely when the techniques are applied to

man. These are not the only areas of research where the principle of arguing by analogy from animals to man may be used. It is possible to employ this principle with regard to the patterns of behaviour.

The study of the functioning of animals is relevant to the study of human function, and the study of their social behaviour is relevant to the study of human social behaviour. There are many studies of the socal behaviour of animals which are used to stand in place of those conducted at the human level. There are many essential investigations of social behaviour which cannot be carried out upon human beings because usually they are too difficult, too dangerous or too time-consuming. There is some danger of facile argument in this and the good behavioural scientist will be on his guard against over-generalization from one species to another. It is also argued later that generalizations should not be a substitute but a starting point for the investigation of the behaviour of man. Be that as it may the study of animal behaviour opens up areas of research which must otherwise remain closed, and allows us to think more coherently about the principles which are fundamental to the behaviour of both animals and man. It allows man to some extent to hold a mirror to his own behaviour and to his own society. It allows him to examine the roots of his social and evolutionary past.

2   Infant Behaviour and the Effects of Early Experience

The anatomist studies the development of an organism by examining the changes which come about in bodily organs from the fertilized egg to the adult animal. Behaviour may also be studied by looking at early forms and by tracing the development of these through to adulthood. In this way it is possible, not only to establish, the principles of behavioural change over the lifetime of the individual, but also to examine the origins of adult behaviour. It could be argued that if we look at the behaviour of adults without a knowledge of the history of behaviour we may never understand fully the principles upon which that behaviour is organized. This is, therefore, an important reason for studying the behaviour of infants. Explanation may be sought in the sequence of change. It is essential to study the behaviour of the infant in order to understand how the behaviour of the adult is formed.

The first task of this chapter is to examine the many forms which infant behaviour takes. The second task is to relate the conditions of social life experienced during infancy to the way the animal behaves at maturity. Early influences are important in changing the animal's character.

The forms of infant behaviour

Social behaviour during the first stages of infant life, at least as far as mammals are concerned, generally takes the form of an interaction with the mother during which the needs of the infant are supplied by her. The mother may be responsible for maintaining the appropriate temperature for the young as well as feeding, washing, retrieving and otherwise looking after them. The infant during this time may be guided initially to the nipple by the actions of the mother but from then on begins an active feeding process. The mother offers warmth and comfort as well as a nourishing supply of food. The young infant's behaviour may be directed towards obtaining these, and the behaviour it shows towards the mother during this time may be governed by these needs.

The first behaviour patterns shown by infant mice are those principally of nursing and sleeping (Williams and Scott, 1953). They emit calls, some of which are well beyond the capacity of the human ear to perceive, that act as a signal to the mother to come and retrieve them should they have become isolated from the litter (Noirot, 1966). The animals begin to groom themselves and to show the face-washing pattern of behaviour which is so characteristic at an early age (Bolles and Woods, 1964). From this point until the time of weaning the animals become considerably more mobile and begin to show fear of strange animals, something which they failed to show before.

Laboratory rats show a rather similar schedule of development. They begin to groom other animals as part of the pattern of social behaviour at or about the thirteenth day of life. Animals which have been maintained in a colony show considerable activity in play. They chase and tumble with one another throughout the colony, using any of the physical features of the environment to assist in this. Playful fighting appears at about the twentieth day of development and serious fighting does not occur until a later stage.

Young pigs establish competition amongst themselves from a very early age. They establish an order of preference for suckling. The most dominant animal attains priority at the teat. This is established by the fact that young pigs develop milk teeth at an early age, which they use to slash at one another in competition for the teat (Hafez, Sumption and Jackway, 1962).

The patterns of play have been carefully observed in puppies (Scott, 1968). It is not always easy to differentiate patterns of play from those which are not. The first apparently playful pattern occurs when one or both pups mouth one another. This activity is more akin to suckling than to biting activity (Rheingold, 1963). The pups then show playful biting, attacking and biting one another in play and also biting parts of the mother's body, particularly her ears and tail. Later the animals play with inanimate objects, such as old slippers, which they will bite and tug. The laces of the observer's shoes were favourite targets for many of these biting attacks.

One pattern of behaviour which then appears and which is widely used in social interactions is one where the animals simply extend one paw to another pup. They frequently fall off balance after this exercise. Later they extend the other paw, giving the appearance of boxing. Tossing the head then enters in as a component in play, and crouching as the introduction to play. Pups would often play by themselves, moving in circles if they should catch sight of their tails or shadow-boxing or generally playing with any inanimate object.

Young monkeys tend to be objects of attraction to the

whole of the rest of the monkey colony (Jay, 1963). Female Indian Langur monkeys in fact allow their infants to be passed from one to another of the females of the colony. At first the most obvious patterns of behaviour are those of nursing and clinging to the mother. The infant confines its activities almost entirely to its mother during the first few months; then she and other mothers leave their infants for a short time and the infant plays several hours of each day with animals of its own age, returning to its mother for nursing. The mother gradually weans the infant until finally she refuses to let it nurse or cling to her. The social behaviour of infants takes many forms. The most significant phases appear to be when the infant first associates almost exclusively with the mother, and when, secondly, it spends a great deal of time with animals of its own age and its behaviour is characterized by patterns of play.

Early experience

We have already seen that the patterns of early behaviour can be important in that they may grow and develop into the behaviour which the adult shows. It also seems to be the case that the kind of life the infant leads can set the standard for the life of the adult which the infant becomes. Early influences which are brought to bear change the shape of future behaviour. One important respect in which change comes about seems to be in relation to the choice of sexual partners. This appears in some cases to be determined, not entirely by the availability or merits of the particular partner, but also by the type of social interaction during early infancy experienced by the animal making the choice of mate. The type of parents, for example, and the experiences of the young with them are very important in this respect.

It may seem strange that an animal should choose a mate of a particular type because of the fact that it has received certain kinds of parentage, but there is evidence for this from animal experimentation. In the choice of foster parents to bring up human infants that have lost their parents or are otherwise separated from them, it is usually the case that care is taken to match the infants as closely as possible to the prospective parents, particularly with regard to physical characteristics. The adoption agency would generally, for example, prefer to place a coloured child with coloured parents and a white child with white parents, if suitable foster parents are available. In animal experimentation it is possible to foster out young animals to parents which are widely different from themselves; in this way the effects of fostering can be assessed and the parental influences observed on the behaviour of the young.

Experiments have been conducted with animals which make use of their different physical characteristics. Pigeons are useful in this respect because they show wide colour ranges. One experiment conducted with pigeons involved different-coloured parents in rearing the same coloured infants (Warriner, Lemmon and Ray, 1963). It is possible to donate the eggs of one strain of pigeons to those of another strain to observe the predetermining effect of early experience. Half the black infants in this experiment were reared by black parents. The other half of the black infants were reared by white parents. When the infants were 40 days old they were isolated, not only from each other, but also from their foster parents, and they remained isolated until they reached sexual maturity. At that time they were placed with other pigeons and the process of choosing a mate began. The male pigeons that were themselves reared by black parents, by and large chose black females with which to mate. This choice provided no test of the views expressed here, however, because black pigeons may choose black mates for genetic or a variety of other reasons unconnected with early experience. When, however, the results were considered of those black male pigeons which had been reared by white parents, it was found that these animals showed a very clear preference for white pigeons as their sexual partners. When male white pigeons were reared by either black or white parents these pigeons also chose mates which were the same colour as their parents. Black or white females, on the other hand, showed no particular preference for a black or white male with which to mate, even though they had been reared by parents of a particular colour. The effect of early experience in this case relates only to the male, who anyway seems to be the active partner in determining the sexual choice.

There are many anecdotes and incidental observations concerning the sexual behaviour of animals reared either by man or by species other than their own. Hand-reared animals may on occasion fail to mate successfully with a partner of their own species. Turkeys, for example, when hand-reared, often appear to prefer human beings as companions to members of their own species (Schein, 1963). These anecdotes and observations need systematic investigations to establish the extent of the transfer of interest from their own to the other species. It may be that a proportion of animals fail to establish appropriate sexual relations even though they have been reared adequately. It may be that a hand-reared bird which fails to establish sexual relations with a member of its own species does so, not through the fact of being reared by a member of another species, but for different reasons. Systematic observations are very much needed in this important area of social development. Observations have been made on mallards. Drakes reared by foster mothers of different species do not always choose partners of the foster parents' species, although a proportion of them do so (Schutz, 1963a). Mallard ducks, on the other hand, almost invariably mate with members of their own species even though they may have been fostered by animals of a different species (Schutz, 1963b).

It is possible to make young animals sexually precocious by the injection of sex hormones (Bambridge, 1962). If chicks have been exposed to a coloured model, so that they have become familiar with it and begin to follow it about, then they direct their prematurely induced sexual behaviour towards this coloured model. Lorenz (1935) supposed that it was through these processes of early response that sexual preferences became fixed. He regarded the process as irreversible. The bond, when once established, does not, however, appear to be generally as fixed as Lorenz supposed. A large proportion of animals when reared with foster parents can often be induced to mate with members of their own species, either when confined with them or exposed to them as a matter of course on a later occasion.

The fact that sexual preferences can be predetermined, and that animals may choose sexual partners from species other than their own, points to the importance of various forms of early learning in regulating and controlling aspects of adult social behaviour. It is important not only to know how far early learning of this kind influences social behaviour, but also how far the effects when once laid down may be changed and modified by subsequent experience.

Investigations of this kind also point to the period of

infancy as one of special sensitivity to external influences – a period during which behaviour and reactions may be either stamped in or modified in such a fashion that the organism never afterwards subsequently responds in quite the same way; a period in which social preferences could be influenced as they are never likely to be influenced again.

Imprinting

One of the important influences which experience brings to bear upon the young animals is that of imprinting. Lorenz (1935) regarded this process as the formation of a bond or the stamping in of an object choice. The term was used initially, with reference to birds, to mean the development of the social attachment between the chick and the mother hen – but it is now commonly regarded as a form of early learning. The chick shows a variety of different forms of behaviour that it directs towards objects to which it has been imprinted. The behaviour appears now to be 'fixed' and, instead of being displayed to a wide range of objects, it centres upon one, or those few objects to which the animal has been exposed. Simply exposing animals to objects having certain characteristics is sufficient to bring about the process of imprinting. Lorenz claimed that imprinting is confined to short well-defined periods early in the life of the chick, that it is irreversible, and that it occurs with considerable rapidity. Lorenz supposes that it is through imprinting that the individual learns to recognize individuals of its own species. Although the claims which Lorenz made appear to be exaggerated, in that, for example, imprinting having once occurred can be made to occur to a variety of other objects, and in that the periods during which imprinting can occur are not as closely defined as was at first believed, nevertheless preferences do appear to be established at an early age and the range of object choice appears to be restricted by this process. Investigations on imprinting have been carried out primarily on chicks of various bird species, particularly those of domestic hens, ducks and geese, because these animals show this phenomenon to a striking degree.

Some of the earliest investigations of imprinting concerned themselves with the response of the young animal, usually a chick or a duckling, to human beings. The response which the chick makes normally to the hen, of approaching it and of continuing to follow it around in a close and intimate fashion, can also be evoked by human beings if they have been responsible for rearing the animals or have had close contact with them during the first few days of their life. Goslings are found to be particularly good at imprinting. If they are removed from the incubator by human beings they will follow this foster parent, persistently peeping in a distressed manner when their adopted parent leaves them (Heinroth, 1911). The original investigators may have felt flattered that these young animals devoted so much attention to them, but they need not, because this filial response, as it has been called, the response of approaching and persistently following, may occur to almost any object. It is as though the chick is incapable of distinguishing, not only human keepers, but a variety of other objects from its natural mother. It must be remembered of course that the young animal on hatching only has a restricted experience of the world, and that these cases where imprinting takes place to objects other than the natural mother usually occur in her complete absence. Imprinting has been found to occur to wooden models constructed to resemble the original parents, cardboard boxes, football bladders, balloons, model railway trucks and matchboxes, amongst many other objects.

Young birds have been found to follow objects of a variety of different colours. It has been reported that some colours are more effective than others in promoting imprinting (Schaefer and Hess, 1959). The results of a variety of investigations are, however, somewhat equivocal on this point. Flickering light sources have also been found to be most effective in the production of the response (James, 1959). Sounds, too, are effective in inducing the young animal to approach them and to explore their origins (Collias and Collias, 1956). The most effective sounds in causing young animals to approach are those in a range lower than 400 K Hz and those which are intermittent and of brief duration (Collias and Joos, 1953).

Imprinting has been studied by the use of operant techniques. Peterson (1960) imprinted ducklings to a moving coloured object. After the initial training the ducklings were required to peck a key in order to view the imprinting object. The animals learnt to peck this key, and the object to which they had been imprinted provided them with sufficient reinforcement to maintain this form of behaviour. That the activities of previously imprinted birds can be controlled by the presentation to them of the imprinting object has been confirmed in other investigations (Hoffman *et al.*, 1966; Cambell and Pickleman, 1961; Bateson and Reese, 1968; Hoffman *et al.*, 1969). The withdrawal of an object to which the animal has been previously imprinted could also exercise control over behaviour. If the object was withdrawn each time the animal approached it, after a while approach behaviour was extinguished. Operant methods can be used in this way to provide a means of investigating the effects of early experience in infant behaviour.

Although the period of imprinting may be extended through particular experimental procedures it is generally regarded as occurring principally during the first few days of life after hatching. The tendency to approach a strange object gives place to the tendency to flee from it during the first three to four days after hatching. The ability to acquire the approach response to new objects declines during the first few days of life. It has been suggested by many authors that there is a critical period for imprinting. This view suggests that there is a limited period of time during which it occurs and that it will not occur if this period has been passed. The term 'critical period' suggests that the boundaries between the time of occurrence and non-occurrence are sharply defined. This is not always the case, however, and so the term 'sensitive period' is usually substituted.

If groups of ducks are reared socially then they appear to become imprinted on one another and thus the sensitive period is brought to an end much more quickly than in animals reared in isolation (Guiton, 1959). Animals imprinted on their companions are no longer free to become imprinted on other animals. It could be the case that imprinting is terminated when once it has occurred as in the previous example, and that the very fact of its occurrence brings the readiness of the animal to imprint further to an end. Alternatively imprinting may finish because other patterns of behaviour which are incompatible with it now appear in the repertoire of the chick's behaviour. It has been suggested that fear is of this form (Dimond, 1968). If the animal shows fear to an object having reached a certain age, then in avoiding the object it cannot show the approach response of imprinting, and thus imprinting is displaced by the variety of behaviour which succeeds it.

Precocial birds are not the only animals to show imprint-ing-like processes. Guinea-pigs appear to follow human foster parents in the same way that young goslings do (Hess, 1959). They also follow moving objects to which they have been previously exposed and keep within a close distance to them (Shipley, 1963). It has also been suggested that imprinting-like processes operate in the attachment of the young to the mother in various primate groups (Harlow, 1961). The question has also arisen as to how far the attach-ment of the young human infant for the mother is an imprinting-like process. It has been suggested that when the young infant smiles this may be analogous to the re-sponses of approach in those animals which show imprint-ing (Gray, 1958). Human infants are necessarily restricted in their movements until quite a late age. Smiling, how-ever, occurs at first in response to a wide variety of objects. Infants may smile at the parents, at complete strangers or at models of faces or even at such undifferentiated stimuli as the heads of mops. The range of objects at which the child smiles is at first a fairly large one, which becomes gradually more restricted by a process which presumably relates to learning, until smiling may occur only at a small number of individuals, or more particularly in response to the presence of the parents. Smiling in response to human figures thus appears to conform to one characteristic of the imprinting process, whereby it occurs at first to a wide variety of objects but becomes gradually more restricted in its scope as the infant develops. It has also been suggested that clinging and sucking are responses which strengthen the bond between mother and infant (Bowlby, 1957) and that these also fall into the category of imprinting-like pro-cesses. It is obviously too early to make extensive claims about imprinting in human beings. It is a difficult research field, but it is one worthy of concentrated effort, and one where important results may await the systematic investigator.

Socialization

The period of infancy is an important one in determining a wide variety of social responses. It is possible that these may be disturbed if infants are reared under unusual conditions or that they may be enhanced by particular kinds of training during early life. It has been suggested that many animal species undergo a period of socialization early in their infancy in which the young animal becomes attached to the group of animals which not only comprises its own species but which also forms its own family group. It would appear that the animal learns at an early stage the characteristics of its own species and produces a definite response which then predetermines, at least to some extent, its later behaviour towards them.

It is obvious in mammals that the formation of attachments is not simply a one-way affair. In the case of sheep not only does the young lamb become attached to its mother, but so also does she become attached to it. If the young lamb is removed from its mother immediately after birth and brought up by human keepers in the absence of any opportunity to associate with members of its own species, then it tends to show its attachment to human beings rather than to other sheep. A lamb of this kind is rejected by the rest of the flock and will be butted away by the adult females who have lambs of their own. Animals of this kind may never be completely integrated into the flock. The social structure and signs of the flock presumably differ from those the young lamb has acquired, and these act effectively to prevent it being given entry. If, however, the lamb is separated from the mother immediately after birth, and then returned to her within the space of a few hours, she will accept the lamb. If it is returned subsequent to that time the mother may reject the lamb completely, butting it away although it is her own offspring.

Mothers of one species may develop attachments for the infants of other species. Sheep exposed to kids immediately after giving birth to their own lambs developed an attachment for the kids, and similarly goats exposed to lambs immediately after giving birth to their own kids developed an attachment towards the lambs (Hersher, Richmond and Moore, 1963). The female sheep appears to establish an attachment to a particular lamb through licking it for a period of 20 to 30 minutes (Smith, Van-Toller and Boyes, 1966). It is suggested that this is the process by which the sheep comes to distinguish her own lamb from others. Separation of the lamb from the mother for periods of up to eight hours does not appear to prejudice the physical development or survival of the lamb, but social and physical disturbance of young goats by the presence of intruders at or immediately after the birth, appears to distort the subsequent attachment the mother displays towards the young (Blauvelt, 1955).

The significance of early periods of development in the process of socialization has been further shown with puppies. Puppies raised by hand from birth show little if any fear of human beings, whereas animals exposed to minimal human contact during the first three months, never form the same relationship with human beings. They remain somewhat timid and aloof. A period between the third and seventh week appears to be particularly important as a time when each animal develops the attachments to members of its own species. The period of infancy is one of critical importance in the development of social relationships and is a time during which experience may radically alter the animal's behaviour.

The effects of isolation and stimulation during infancy

The opportunity to learn the characteristics of one's own species represents only a small part of what is implied by the term socialization. Social environments afford greater oportunities than that. If the adults of many species are isolated from contact with other members of their own species as well as members of other species, they may develop behaviour disorders which become obvious both before and after they have been reintroduced to companions. The process of socialization is a continuing one, and one which is relevant to adulthood. The period of infancy, however, seems to be particularly important as a time when social influences of all kinds exert their maximal effect and as a time at which they have the greatest potential for bringing about change. Puppies are found to act like wild animals if they are fed without contact with human beings (Scott, 1968). The well-known experiments of the Harlows (1965) must be mentioned in this context. They found that young rhesus monkeys are unable to establish satisfactory social adjustment if they have experienced social deprivation between the third and the sixth month of life, and have during that time been prevented from making attachments between themselves.

Many young animals cannot survive the earliest periods of life without considerable care and protection. Rats are born blind and deaf as well as lacking in hair. They are unable at first to urinate or defecate spontaneously. The effects of stimulation on young rats and mice are marked. If they are taken by the experimenter in his hand and held there for a short period of time and then returned to the nest, the behaviour of these animals as they grow up will be different. Rats treated in this way during infancy grow faster and live longer. Their behaviour is less emotionally disturbed when they are subsequently placed in a strange situation, and some reports suggest that animals handled in this way withstand stress more adequately (Denenberg, 1967). The results point to the importance of various forms of stimulation during infancy, and raise interesting questions as to how far socialization is carried out by the mother herself and how far the interaction with the other members of the litter contributes to this process. The argument that young rats could be regarded as being partly socialized by human beings, and thus appear less disturbed in situations in association with them, cannot be maintained, because a variety of different kinds of stimulation unrelated to direct human stimulation produces similar effects. The suggestion arises that early stimulation modifies the development of the adrenal glands, acting to change the way in which they function on subsequent occasions.

Many experiments show that animals living in barren environments are deficient in learning performance. It is known, for example, that chaffinches reared in isolation develop only a poor substitute for the usual chaffinch song (Thorpe, 1965). It is a restricted sub-song. Birds reared with parents develop the full song with all its elaborations, and birds reared together as groups of young develop somewhat the same kind of song, but one which has elements which are peculiar to the group. Birds reared in isolation are prevented from developing to the full extent the song, which affords an important means of communication. It is possible that restriction in this fashion denies the animal access to all kinds of social response and behaviour which may be important, not only in deciphering the response of the community of animals towards it, but also in making known its own facility for interaction, and its own individual needs which relate to the behaviour of others.

There appears to be some evidence that stimulation during infancy acts to enlarge brain size (Bennett *et al.*, 1964). Stimulation may take the form, for example, in an experiment on rats of providing them with objects in the cage, e.g., wooden bricks, with which they can play and become familiar. There is no doubt that early stimulation also acts in the vast majority of cases to improve the animal's ability to perform various learning tasks (Forgays and Read, 1962). What has been described as the plasticity of behaviour has somehow been increased. The animal is capable of responding in a much more varied and subtle fashion. Not only may the process of socialization serve to teach infant animals the modes of social response and the language of behaviour which they must use to communicate with other animals, but socialization may also have other effects in perhaps increasing the plasticity and modifiability of behaviour, such that it becomes more readily adapted and changed to suit the behaviour of the companions with which the animal finds itself.

The distinction between innate and learned behaviour

Finally, we have to consider one of the most troublesome and difficult areas of all in the study of infant behaviour. The arguments for and against the division between innate and learned behaviour are presented here briefly. They are of particular concern to the study of neonate and infant behaviour.

First, it is commonly held that there are patterns of behaviour which can be described as innate. These are distinguished because they may occur in a stereotyped form throughout representative members of an animal species. This suggests that the behaviour is under the influence of hereditarily controlled mechanisms. Secondly, particular patterns of response may appear in neonates isolated from members of their own species. These animals, it is argued, have been denied the opportunity of learning this behaviour from companions.

The problem of instinct has been a recurrent theme in the history of the study of animal behaviour. Just as there are arguments for innate behaviour, so there are arguments against it. First, the use of the term innate may be a description rather than an explanation. It does not help to call behaviour innate if by so doing the need for further analysis is ignored. Secondly, negative definitions are not helpful in this context. It is not productive to define innate behaviour as that which is not learned, nor for that matter is it useful to define learned behaviour as that which is not innate. The definition of what is learned is as imprecise as that of what is innate. A negative definition leads to a 'rag bag' collection of different forms of behaviour, each in their own right demanding some form of explanation. Thirdly, it has been argued that the neonate animal is by no means free from environmental influences at birth, whether the environment is the external world or that of the internal world of the mother (Lehrman, 1954; Dimond, 1970). The opportunity for learning exists throughout the embryonic period as well as during neonate development. Learning need not be represented by an attempt at deliberate training but can take the form of perceptual registration in which the organism acquires information about the environment without necessarily seeking reward (Sluckin, 1964). Social isolation of an ani-

mal from companions by no means ensures that it is free
from influences which bring about learning.

Hebb (1949) has taken a rather different direction in
this controversy. He supposes that the argument as to
whether behaviour is innate or learned is largely sterile
because behaviour is the product of both learning and
heredity. Learning necessarily contributes 100 per cent
and heredity necessarily contributes 100 per cent. Whilst
agreeing with Hebb that the attempt to distinguish between innate and learned behaviour is essentially sterile,
it cannot be agreed that the contribution of learning and
heredity to behavioural processes should be described in
terms of a percentage. Hebb's statement presupposes an
analysis of this kind.

It is surely not possible to give weighted scores to
heredity and learning, and the idea that each can be separated out from the other as a fraction of behaviour appears
to be misleading. It fails to make sense to attempt to attribute the reason for the smoke coming from a fire to the
material which is now burning or to the hand which
applied the match. Analysis in these terms is inappropriate
to the nature of the problem being studied.

One way forward seems to be to concentrate on the
development of behaviour itself rather than attempt to
discriminate between innate and learned components. It
is necessary now to understand the development of behaviour over time. Many of the previous difficulties arise from
considering behaviour over only a short time-span and not
in an historical context. To understand behaviour fully
we must examine the extended span of behaviour from the
earliest origins to its most sophisticated development. We
cannot adequately attribute causes to behaviour without
studying development, and the history of the changes
which come about in behaviour are surely the sources for

our explanation of behaviour. To attribute causes to behaviour without studying the development of that behaviour is to guess in the dark. Development is itself a rich
source of the explanation and principles we are seeking.

To this end behaviour can be described in terms of the
developmental complex. This is an identifiable unit of
behaviour. This is behaviour which can be analysed in
terms of movement patterns, but ones nonetheless which
can, and possibly do, change over time as different influences come to bear upon them. The developmental complex is the product of many influences including those of
the pre-existing substrate, the mechanisms of heredity, as
well as the organism's response to the events of the environment, and the registration of these events upon the organism.

There is no real substitute for the investigation of the
history of behaviour. There is no real substitute for the
endeavour of understanding development by meticulous
research. There are, however, rich sources of hypotheses in
the study of development itself, and the developmental
complex is a means of analysing these problems to get a
better understanding of them.

Summary

The importance of the study of development in understanding behaviour has been stressed in this chapter. 'The
child is the father to the man', and so we need to study the
behaviour of the infant to see how the behaviour of the
adult develops out of it.

Animals typically show different patterns of behaviour
at different ages. The first task was to examine the nature
of these changes to the conditions of life experienced dur-

ing infancy, to the way the animal behaves on reaching maturity.

The kind of life the infant leads can set the standard for the life of the adult which the infant becomes. Early experience changes and moulds the behaviour of the organism. The choice of sexual partner, for example, appears to some extent to be predetermined by the nature of the experience of companions and parents when the organism was young. This work points to infancy as a time of special sensitivity to external influences as a time where important reactions may be stamped in or modified in such a way that the organism never responds subsequently in quite the same manner.

One of the illustrations of this is in the study of imprinting, whereby simply exposing certain animals during early infancy to objects having particular characteristics is sufficient to predetermine their choice for these objects on later occasions. Imprinting has been studied by the use of operant techniques. Ducklings will peck at particular keys in order to reinstate the presence of an object to which they had been previously imprinted. The activities of previously imprinted birds can be controlled by presenting them with the imprinting object as a reinforcement. Operant techniques have been used successfully in a variety of investigations to study the effects of early experience. The use of these techniques points to a view of learning as perceptual registration.

The period of infancy is important as one in which socialization occurs. Socialization may be disturbed if infants are reared under unusual conditions or it may be enhanced by particular kinds of training. Animals isolated during infancy may develop a variety of behaviour disorders. This is marked in the development of the social behaviour of primates. It is suggested that the process of socialization is the means by which animals learn the modes of social response which they use to communicate with other animals. If they are denied this opportunity for socialization, they are prevented from learning the behavioural vocabulary or dialect and the understanding of this which is used to bring about social interaction.

The study of infant behaviour is a touchstone in the argument as to the innate or learned aspect of response. The view is expressed here that this is a sterile controversy, and it is suggested that emphasis should be switched from the attempt to allocate fractions of behaviour into innate or learned compartments to a thoroughgoing study of development itself, using the developmental complex as the unit necessary for the analysis of behaviour.

3   Parental Behaviour

This chapter directs attention to parental behaviour, not only in respect of the marked influence which it has upon the young, but more particularly to try to understand its organization and the reasons why it takes the form that it does. It is essential to remain aware that the behaviour of the parent can really only be fully described in relation to that of the infant and, although it is necessary to describe some aspects of infant behaviour, this is secondary, however, to our concern with the behaviour of the parent.

The male and female may share the responsibility for the care and protection of the young. In some species the male may bear the major responsibility for parental care, but this is unusual, and certainly as far as the mammals are concerned and particularly during the early stages, the female is most frequently entrusted with the care of the young. Animals which are not the parents may also be involved in looking after and caring for the young, and systems of community care may develop in some cases in which the whole colony or group of animals may not only show tolerance but also interest and protection towards the young (De Vore, 1963).

The young of advanced animal species are often born in an utterly helpless condition. They depend on parental care for their very survival. If this should be withheld through accident or for some other reason involving the loss of the parents, then the infant is unable to defend itself and dies through exposure, cold or hunger, or becomes a prey to other animals. Parental behaviour is of considerable evolutionary significance. Those animals which are able to look after their young not only ensure the survival of the young but also the survival of the race. This is shown most clearly in the case of inadequate parental care. Deficient mothering may lead to the death or permanent weakening or disablement of the offspring. The offspring of inadequate parents may, if they survive, also be so disturbed that they develop serious social deficiencies on their own account which handicap them in later life. These may take the form of an incapacity to mate or rear infants of their own. This effect could make it impossible for the animals to breed, with the consequence that the lineage which they represent has a small chance of survival. Genetic transmission of inadequate patterns of parental behaviour becomes impossible if the offspring of bad parents fail to survive. On the other hand, those animals which show the very best parental care at least create the conditions for survival, and since the offspring are unlikely to be burdened with the social disabilities described earlier they will reproduce normally and continue their lineage. In evolutionary terms there is strong selection pressure against thoroughly inadequate parental behaviour, leading to its eradication in one or at most two generations.

The parent in many animal species is responsible as an agent for guiding and for predetermining infant behaviour. The parent, in punishing or rewarding the behaviour of the young, exerts authority and acts as a means of changing their behaviour. Particular patterns of behaviour in the repertoire of the infant can be strengthened or extinguished by the behaviour of the parent. The parent can act as a positive or negative reinforcing agent. The parent not only develops appropriate skills in the infant through patterns of play and social stimulation, but also provides important constraints with which the infant becomes familiar at an early age. Constraints can set the limits to acceptable behaviour and it is through these, coupled with the many positive reinforcing situations, that parental control is exercised.

The extent to which the infant imitates the behaviour of its parents is still open to question. Parents may teach their offspring largely through their own example, and the infant may actively copy the forms of behaviour to which it has been exposed. Although learning could occur because the infant copies the behaviour of its parents, there are still important constraints and rewards in this situation which could themselves lead the infant to perform this copying function rather than to behave in other possible ways.

The study of the interrelation of the infant with the parent in animal species is, therefore, important in a variety of ways, and the study of parental behaviour in animals is also of value in that it highlights important areas for investigation in human beings. It acts as an analogue of the processes governing the care of human infants and points to important areas for investigation at this level. We are limited in the type and nature of the information which can be gathered about parental behaviour in human beings particularly with regard to its biological determinants, and many of the important questions we should like to ask about human beings are not yet open to experimental investigation. Studies of parental behaviour in animals extend our knowledge into these regions and evidence obtained from non-human species must here stand in place of that gathered directly from experiments on man.

Pioneering work in the understanding of parental behaviour was carried out by Sturman-Hulbe and Stone (1929) and by Wiesner and Sheard (1933). These early investigations did not receive the full attention which they merited at the time. It is only in recent years with the growth of interest and an appreciation of the significance of the problems of parental behaviour that full information is now becoming available to us.

Phases of parental response

One of the most important of the responsibilities which the mother bears towards the young is that for feeding and for supplying them with nutritive substances. In mammal species this feeding relationship can be described as the nursing phase (Rosenblatt and Lehrman, 1963). The nursing phase can be divided into three parts.

In the first phase the female plays the major part in initiating feeding. The young may perform nuzzling or sucking movements but the female at this stage is almost solely responsible for guiding the infant to the nipple region. She stands over the young ensuring that the nipples are within easy reach in order that they may feed. The female by her behaviour directs the nipples to the young and facilitates their feeding in every possible way. This phase may last for different periods of time in different species but it is certainly one which is dominated by the female and in which she plays a very active part.

In the second phase of feeding the infant takes a much more active part. The infant may initiate feeding, actively seeking and demanding the female's attention. The female readily assists feeding by remaining in the nursing position and by making the nipples readily available.

Finally, in the third nursing stage the infants may continue to feed from the female, but as time goes on she becomes increasingly reluctant to assist them in feeding and removes the nipple region from them. Weaning is one of the last of the maternal relationships, and it is one in which the female takes active steps to prevent the infant from feeding from her.

The maternal behaviour of most mammals takes this phasic form. It is uncommon for virgin females to show a parental response. When the female has given birth she looks after the young until they are old enough to look after themselves, and then she withholds maternal care from them.

The phase of maternal behaviour in mice was measured by King (1963). He found that the females spend an increasing amount of time looking after the young in the nest during the first few days after their birth, but from then on the time spent with the young declined until weaning. The females spent less and less time nursing as the animals became older. The females became much more defensive when they had young in the nest. Pregnant mice could be easily chased away from the nest, but after the birth of infants the mothers savagely defended the nest area. Strange males are vigorously repulsed from the nest area, but as the young become older so the ferocity of the female's attack on strange males diminishes. When the infants are 16 days old, strange males are seldom attacked at all.

The parental cycle follows very much the same course in cats (Schneirla, Rosenblatt and Tobach, 1963) and in dogs (Rheingold, 1963).

The changes and development of maternal behaviour in rhesus monkeys has been fully described (Harlow, 1960; Hansen, 1962; Harlow, Harlow and Hansen, 1963). This proceeds through three stages: (1) attachment and protection; (2) ambivalence; and (3) separation and rejection.

In the first stage the infant clings to the ventral surface of the mother and is cradled by her arms and legs. The infant spends much time at the breast. During this period the infant not only spends time in nutritive sucking but also maintains contact with the breast without sucking. Grooming of the infant by the female increases steadily through the first two months and this could represent the strengthening of the parent–infant bond.

During the second stage of development the infant becomes much more active and the female spends time in restraining it. The infant tries to break away from the mother, who is now forced to become much more vigilant and actively engaged on its behalf. Ventral contact, non-nutritive sucking and grooming responses all show a gradual decline during this period. Hinde and Spencer-Booth (1967) examined the parent–infant relationship in socially living rhesus monkeys. They showed that the period of time spent on the mother decreased as the infant became more mature. The time spent at the nipple increased during the first few weeks but declined thereafter. Older animals may spend long periods of time on the mother but not attached to the nipple. The same authors showed that at first the mother behaves in a restrictive fashion. As the infant moves away from her she moves after it; later, however, the infant approaches the mother more often than she approaches it, but if the infant has been frightened it runs and clings to the mother. Infants in the wild may have relatively greater freedom than those in captive groups. Kaufmann (1966) records that infants in a wild troop may well spend much time out of contact with the mother even during the first month of life.

In the final period of rejection and hostility the mother

comes to punish the infant more and more. Vigorous biting of the nipples by the infant and tugging at the mother's hair could cause her pain, leading her to be less attached to the infant than before. This could be one factor responsible for the waning of maternal response. The hostile behaviour of the mother could lead to the separation of the infant from her. Immanishi (1957) reports that in the feral state most males leave by the end of the second year except those with especially dominant mothers, but female Japanese macaques may remain close to their own mother although they may be mothers themselves. Hinde and Spencer-Booth (1967) point out that some components of maternal rejection occur during the first week of the infant's life, and they stress the role of the mother in promoting the independence of the infant. The young infant may be restrained and prevented from leaving the mother, but the older infant may be punished for approaching her. This suggests that the mother effectively terminates her contact with the young, and it is she who is responsible for dissolving the relationship between the parent and the young.

Measures of parental response

There are a number of measures which have been used to provide an index of the strength and intensity of parental response; principal amongst these are retrieval, nursing, grooming and nesting. The use of nursing as a measure of this has already been described.

   In many species the mothers not only restrain their offspring but also actively retrieve them when they stray from the nest or home area. Beach and Jaynes (1956a) showed that female rats will retrieve a variety of pups during the first few days after parturition, and retrieval is not at this stage confined to the animal's own young. As time goes on the female becomes more selective: when faced with a choice, she now retrieves her own pups before those of an alien species or those from another litter. If young mice are removed from the nest the female emerges from the nest to retrieve them by picking them up in her mouth and removing them bodily back to the nest (King, 1958). On the first day after giving birth to the young the mother is not very good at retrieving them. She may at this stage attempt perfunctorily to build nests over the young where-ever they happen to be. Later on she improves consider-ably, retrieving the first animal and then emerging from the nest to retrieve the others.

Female monkeys also retrieve their infants. Hansen sug-gests that the females may use a special code in this respect. When the infant had strayed to a remote place and the mother appeared particularly alarmed, she would show behaviour which could be described as a 'silly grin'. This is a facial grimace which occurs rarely but is remarkably effective in promoting the return of the infant to the mother. Hansen suggests that there are a variety of signal systems between the mother and the young of this type. Hinde and Spencer-Booth (1966) suggest, however, with regard to this particular response that it is not of necessity a means of reuniting the infant with the parent but that it may be a fear response on the part of the female to the separation which now exists between the infant and her-self.

Retrieval is clearly related to the transport and protec-tion of the young under normal conditions. As a measure of parental response, however, it is complicated by the independent activity of the young. The young may assist their retrieval by running to the mother of their own accord, or they may attempt to avoid the attempts of the mother to reinstate them in the nest.

Grooming is one important pattern of parental behaviour relevant to the formation of the bond between the mother and the infant which can be used as a measure of the strength of the maternal response. Jay (1963) reports that the young of the female langur monkey are groomed from the moment they are born. When the infant is asleep it is groomed and stroked without being woken. The infant struggles when it is held away from the mother's body for inspection and grooming, but the female remains in complete control on these occasions. If the infant should cry, then again the female grooms it. The more the infant cries, the more frequently it is groomed.

De Vore (1963) describes the treatment administered to infant baboons. These receive the mother's undivided attention during the first few weeks of life. Frequently the mother licks and nuzzles the infant and explores its body.

This behaviour has biological advantage in removing parasites and in keeping the infant clean, but grooming also has a wider social significance. Grooming is the means by which social status amongst the group is indicated and is possibly a means of establishing and maintaining pair relationships. Possibly the wider social significance of this lies in the maternal relationship. Older females may be allowed to groom the infant after the first few weeks, but only whilst it remains in the mother's arms. Older males may aproach and touch the infant. They may also have responsibility for carrying infants upon their bellies during long treks. The adult males are highly sensitive to the distress of the infant, and they may well attack a human observer, should he come between an isolated infant and the rest of the group. Other animals are frequently gathered in grooming clusters around mothers with young infants. Grooming appears to indicate friendliness and goodwill.

Grooming appears to perform valuable social functions as well as those directly concerned with the care of the young. Most commonly the mother takes the major responsibility for grooming her own infant but, as we have seen, other animals may also assist in this aspect of parental care.

Nest building is a pattern of behaviour which occurs often in anticipation of the needs of the young. It serves an invaluable social function in acting to protect both the mother and her young. Rats build nests during the latter stages of pregnancy. Nests are built out of a variety of materials. The method of building and construction depends to a large extent upon the nature of the material which is used. As the material accumulates at the nest site the female spends less time in carrying it there and more time in shaping the existing material. Nest building and shaping continue for two weeks after the birth of the young, and then declines. The pups no longer huddle at the nest site as they become active, the nest as such becomes flattened and the female no longer continues to attempt to build it up around the young (Eibl-Eibesfeldt, 1955).

Female rabbits use their own hair as nesting material with which to line the nest (Swain, *et al.*, 1960). The hair becomes loose at the time of late pregnancy. This can be measured by passing a comb a specified number of times over a particular area of the body and by weighing the accumulated hair. Hair loosening reaches a peak at the time of parturition, and this is one more stage in preparation for parenthood which is presumably related to the endogenous changes which pregnancy brings about. When the female rats are allowed to keep their own young, nest building continues during the first two weeks after their birth, and is absent from the eighteenth day onwards. If, however, the young have been separated from the female at parturition, nest building declines almost immediately (Rosenblatt and Lehrman, 1963). It is an activity which is closely related to the needs and physical presence of the young. It cannot be argued, however, that the physical presence of the young is solely responsible for nest building, because in many species the female constructs a nest before giving birth to the young. It seems to be an activity associated with the endogenous changes of late pregnancy. It is possible that these changes, when once they have been set in operation, are in some way maintained by the stimulus which the presence of the young provides. In this way the young could themselves be responsible for maintaining the behaviour which is appropriate to their care, but initially the female may build the nest before the young are born. Nest building may relate to the care and protection of the female during late pregnancy as well as to that of the young when they are born.

The effect of additional young

Parental behaviour has been studied by experimental manipulation as well as by observation. One technique is to present the parents with young mice which are not their own. Maternal response to these additional or substitute pups usually makes an appearance at, or shortly before, the parturition of the animal's own first-born young, from which point on it persists until the animals are old enough to look after themselves. In most animal species it is uncommon for virgin animals to show a marked parental response to the young. Richards (1966a) found that virgin female hamsters, far from showing maternal response to young test pups presented to them, turned on them and killed them as though they were prey. Pups aged from one to six days are frequently attacked and killed by virgin females. Older pups may be treated with a mixture of attack and maternal response, whereas older animals still may be treated in a fashion similar to any strange adult intruder (Richards, 1966b). Pregnant females, on the other hand, combined initial attack upon the young with subsequent maternal response towards them.

Rosenblatt (1965) describes studies in infant–mother interaction in the laboratory rat, using this technique. The mothers were allowed to keep their own litters, and pups aged five to ten days were presented to them at different stages in the maternal cycle. The mothers at and shortly after birth showed the full maternal response to these test young. They retrieved them, took them into the nest and licked, nursed and groomed them. Later on in the cycle the female behaved in a far less motherly fashion towards them, and during the fourteenth to twenty-fourth days after parturition her maternal behaviour declined.

Females which had not yet given birth to their young were tested. Nursing and retrieving behaviour was not shown towards test pups until the female had given birth to her own young, but nest building occurred more and more frequently with the duration of pregnancy. This work suggests that the onset of parental behaviour in laboratory rats is associated with the late stages of pregnancy and the process of giving birth to the young. The work also suggests that there are endogenous changes which come about in the mother herself which lead ultimately to the waning of maternal response.

It is difficult, however, to assess the effect of the female's own maturing young upon her maternal behaviour. It became necessary, therefore, to separate out these effects and to try to distinguish them from the endogenous changes which take place on the part of the mother. Rosenblatt and Lehrman (1963) carried out experiments in removing the animal's own young and replacing them with foster young. The foster young remained with the female for a short while before they were themselves replaced with a fresh group of young. In this way Rosenblatt and Lehrman sought continuously to provide the females throughout the maternal cycle with young whose age varied only within a narrow closely prescribed range. Maternal reaction to these foster animals again declined rapidly throughout the maternal cycle. This evidence suggests some endogenous decline in maternal response, although the interpretation of the results of this experiment is complicated by the fact that, in order to maintain the maternal state, the animal's own young were returned to her between periods of foster care.

Maternal behaviour declines very rapidly in the total absence of the stimulus provided by the infant. Rosenblatt and Lehrman (1963), in a further investigation of maternal response in the rat, separated the pups from the mother at parturition and returned them to her after different intervals. They showed that the response of the female declines sharply as the period of absence is increased. After a few days of separation the female no longer responds in a particularly maternal fashion although she has been reunited with her own pups. The bond which would otherwise have been formed between the infant and the mother no longer exists.

These studies point to three important factors in the organisation and control of maternal response. These are: first, the changes resulting in the institution of maternal response associated with late pregnancy and parturition; second, the effect of the stimulus provided by the pup in promoting and maintaining the response; third, those endogenous factors, as yet unspecified, on the part of the female which lead to the waning of the response.

The effects of experience

The question arises frequently as to how far experience in the animal's past history contributes to parental behaviour. Would it be possible, for example, for animals, completely isolated from others and deprived of any form of social contact, to show maternal behaviour and display the patterns of behaviour which are necessary for bringing up the young?

It is reported that animals which have been prevented from licking themselves for a period of time prior to parturition show disturbed maternal behaviour. Birch (1956) placed rubber collars around the animals' necks which effectively prevented self-licking. This severely reduced the subsequent efficiency of the mothers in caring for their young. The evidence with respect to the importance of self-licking as a contributory factor to maternal response appears, however, to be in some doubt. It is possible to require one group of animals to wear collars which will interfere with self-licking during pregnancy, and another group to wear collars which will not. Animals in both cases have been found to deliver and maintain a litter satisfactorily in ways which did not differ significantly from one to another (Christopherson and Wagman, 1965).

Eibl-Eibesfeldt (1958) also cites a study by Coomans in which it is suggested that any differences in maternal behaviour could be attributed to the interference which a collar may produce in the normal patterns of behaviour. Although self-licking may not be as important with regard to subsequent behaviour as it at first seemed, the question still arises as to how the female comes to perform in the way that she does to look after the young.

Animals without previous experience of giving birth or of rearing young are found in many cases to rear their young perfectly adequately. The rate of survival of animals from the first litter may not differ materially from that of subsequent litters. Rats show perfectly adequate care towards their first litter and, if there are any effects of experience, presumably these lead the animal to become more adept and skilful at handling subsequent litters. The skill need not, however, be reflected in any enhanced capacity of the young to survive.

Rabbits show little change in direct maternal care between the first and succeeding litters as measured by interest in the young, aggressive response, and time of nest building in relation to parturition (Denenberg *et al.*, 1958). When, however, less direct measures are used, such as the quality of nest building (Ross *et al.*, 1959), then the standard of construction of the nest has been found to improve over the first three litters, but not over subsequent ones. Cats also appear to care for their first litter in a way which is perfectly adequate and at a standard which is in no way diminished below that given to subsequent litters. On the first litter, cats tend to lick themselves before licking the kittens. On subsequent litters, they lick the kitten before they lick themselves (Schneirla, Rosenblatt and Tobach, 1963), but apart from that there is little in the parental behaviour of the cat which could be pointed to as an improvement in care over subsequent litters. It can not be assumed on the basis of this evidence, however, that experience has no effect whatever upon the level of maternal response. It may be that there are effects but that the rearing of several litters can have no effect over and above that of rearing one litter. It can be stated, however, that there is little in the way of improvement in maternal care in these animals from one litter to the next.

Primates also appear to rear the young adequately after the first parturition, but the effects of having given birth previously are more marked. This is illustrated in the viewing behaviour of the female towards her young. Operant methods have been used to study the development of maternal responsiveness of primates (Cross and Harlow, 1963). Monkeys can be placed in a viewing box which contains a window through which they can look into another cage. The monkey may be required to open this window. It is possible to assess the interest which a monkey shows in objects placed in the other cage by observing the number of times the animal opens the window to look through at them. Social relationships can be investigated by placing one monkey at the other side of the window and by observing the number of times another monkey opens the window to look at it. This viewing method is one which can also be used to assess the social interests of pre-verbal human infants. Female monkeys were allowed to view an infant or a juvenile monkey through the process of opening the window. Females not yet having given birth showed no particular preference for baby viewing. After giving birth they showed a distinct preference for the infant and viewed it far more frequently. A group of females, each having given birth several times, showed an obvious preference for viewing the baby. Contact with the baby seems to be reinforcing to the female. If the female is deprived of contact with the infant the response of viewing the infant gradually declines (Seay *et al.*, 1962). The behaviour of the infant in addition to that of the female may be affected by the question of whether the female has previously given birth or not. First-born infants play less with their parents, show less emotional and self-directed behaviour and appear to be more disturbed by strange environments (Mitchell *et al.*, 1966). Primates, therefore, appear to be affected to a much greater degree by the experience gained over subsequent litters of young than do some lower animals.

Although differences may be observed between the behaviour shown by female monkeys to their first and subsequent litters, these differences cannot in any sense be regarded as an explanation of maternal response. Many animals show perfectly adequate parental behaviour in bringing up their first as well as subsequent litters, and it is still necessary to seek an explanation of this.

The separation of the parent from the young

One of the most important techniques for studying parental behaviour consists of separating the young and the parent from each other and in returning the young to the parent after a variable period of time. The studies of Rosenblatt and Lehrman (1963), in which young laboratory rats were removed from their mother, make it clear that the presence of the young is necessary in order that maternal response may persist in this species beyond the time of parturition. If immediately after giving birth the females were deprived for a period of more than five days of the young to which they had just given birth, then the litters which were given to them subsequently failed to survive. Experiments of this kind emphasize the importance of the period immediately after birth as a time during which maternal behaviour becomes fixed as a more stable and persisting pattern of response. Collias (1956) also emphasized this period as being one of importance

in determining whether maternal response would diminish or persist. New-born kids or lambs separated from
their mothers at or shortly after birth for periods of two
hours or more stand a high chance of being rejected by the
mother when they are subsequently returned to her. There
appear in these species to be at least two clear components
in the establishment of maternal response. The first is
the institution of maternal response. This appears to be
associated with late pregnancy and the process of giving birth. The other factor is the one which causes maternal behaviour to persist when once it has been instituted.
The presence of the young and the stimulus provided by
them is necessary to this second factor.

The loss of the young has a marked effect on the female.
She becomes far from passive. She becomes agitated and
disturbed and may search vainly around the cage. Rhesus
monkeys following the loss of their own young may
attempt to adopt other animals. Harlow, Harlow and Hansen (1963) reported that a rhesus monkey attempted to
adopt a kitten placed in the case with it following the loss
of its own young. The mother retrieved the kitten and
attempted to attach it to her nipple, but she abandoned it
when it failed to cling to her or create ventral contact with
her. Females whose own infants were removed on the day
of birth were found readily to adopt fresh infants of the
same species although these may have been several months
old. Females which had ceased in the absence of infants to
lactate could be persuaded to do so again as normal
mother–infant relationships were resumed. When older
infants were separated from their mothers, an intense relationship was demonstrated when the infants were once
again reunited with their mothers (Kaufman and Rosenblum, 1967). There are many investigations which show
the effect of maternal deprivation upon infant behaviour.

This work is reviewed by Sackett (1967), who suggests that the infants are not the victims of arrested intellectual development but are extremely fearful in the strange test environments which are normally used to investigate intellectual development. Mitchell *et al.* (1967) report that maternal separation early in life leads the animal to become persistently more fearful, and this may continue for periods of over a year after the initial separation. The infants may show withdrawal and experience difficulty in adapting to new situations (Griffen and Harlow, 1966), and they may show depression similar to that of human infants (Kaufman and Rosenblum, 1967).

Just as there are effects upon the infant of separation from the mother, so there are effects upon the behaviour of the mother. The mother when separated from the young begins an active search for them; she shows an increase in activity level which diminishes somewhat over time. If, however, the female can hear the infants or glimpse them occasionally, her movement becomes considerably reduced (Jensen and Tolman, 1962). If a transparent screen divides the mother from her infant, then again this has a marked pacifying effect upon her. She became less active when her own infant was placed in the adjoining cage but continued to pace about when an older infant was placed there (Jensen, 1965).

Separation of the female from the young not only disturbs the behaviour and development of the young but also exerts a profound influence upon the female. Short periods of separation lead to considerable agitation on the part of the female. She gives every appearance of distress and begins a process of active search. In some species short periods of separation are sufficient to bring maternal response to an end.

Problems in the organization of parental behaviour

The development of maternal behaviour follows a definite sequence of events. The parental response to the young consists of a number of different types of behaviour which are all highly correlated with each other. They all appear at very much the same time shortly before or after the birth of the young and diminish gradually with the approaching maturity of the infant.

The principal patterns which are coordinated in this way are those of nest building, retrieval, licking and caring for the young, feeding and maintaining contact with them as well as showing an aggressive response to intruders.

The young develop from the mother and, because of the nature of the feeding relationships, the young before birth are parasitic on the mother; after birth they also could be described as parasitic upon her. The mother is required to nourish the young both before they are born and after they are born. It seems likely that the placental hormones or related substances act to bring about a transfer of energy reserves from a direct supply of nutritive substances through the blood stream to external feeding through the mammary surface. Lactation through the nipples and the stimulation of them by the young may be one of the principal factors in the onset and maintenance of maternal behaviour in mammals. The female certainly plays a large part in guiding the young to the nipple during the early feeding relationship. It could be that the presence of milk or hormones associated with milk production leads the nipple region to become particularly sensitive. The female under these circumstances may not only supply the litter with food because they are demanding it but also because in the post-partum condition she is seeking nipple stimulation.

It is possible that many other patterns of maternal response could be derived secondarily from this feeding relationship. There are, however, other important factors to be considered, and principal amongst these are the protection of the nest area and the need to conserve body temperature. It is commonly supposed, for example, that nest building affords a means of protection to the developing young. It also prevents the young losing heat at a time during which they are essentially unprotected by the hair which develops later upon the body. Whilst the nest obviously has these important functions in relation to the young, nest building as a form of behaviour could be related also to the female's own state. Nest building often occurs in late pregnancy before the birth of the young. It would be hard to argue that the female constructs the nest because she has foresight of the needs of the young which have not yet been born. During the late stage of pregnancy animals frequently become inactive and lethargic. It is important for the inactive animal to conserve heat. Nest-building could function to prevent heat loss from the inactive female during the last few days of pregnancy. It is certainly the case that rodents show greater nest-building activity in lower ambient temperatures, and it may be the case that heat loss resulting from lack of activity is the trigger to this response in the pregnant female. Insofar as the young and the female subsequently share a common thermal environment, then both the female and the young are protected and provided with comfort by the efforts of the female to maintain the nest.

If the female undergoes a stage of relative immobility during late pregnancy, then it is also essential that she should be protected during this time, and that she should escape detection from any possible predators. It is equally important that these conditions should hold after the birth of the young. It is important, therefore, that the female should construct the nest before giving birth and that both she and the infants should remain hidden. Infants escaping from the nest are likely by the nature of distress cries to reveal the source of the nest to predators. If the mother retrieves the infant, then she again safeguards the nest, the litter and herself from danger.

The stage of maternal separation comes about when the mother withdraws her attention from the young. It is possible that maternal response terminates as the result of endogenous change within the female. There is certainly evidence to suggest that maternal response cannot be maintained indefinitely. On the other hand, it is known that maternal response in a variety of species may rapidly terminate if the young are removed at birth. This suggests that the young themselves are not without influence in leading to the perpetuation of maternal response, and it may be that the young are also ultimately responsible for the termination of maternal behaviour. It is likely that, as the young mature, the female's encounters with them may be unpleasant rather than rewarding. The female will now try to avoid the infant and remove herself from contact with it wherever possible. The female could under these circumstances still retain her response to the younger animal, whilst at the same time taking care to avoid older infants.

The increase in size of the young as they become older could be another factor in the termination of maternal response. Older young may show behaviour patterns which closely approach those of the adult. Older animals are, of course, much larger than younger ones; they may not only be more vigorous and strong, but are more sophisticated in the range of behaviour available to them. In many cases the older generation finds itself in competition with the younger, to which at some stage it inevitably has to give place. The older parent, therefore, often has genuine reason to fear the younger animal as it approaches maturity, and it may well be that the assumption of adult behaviour on the part of the infant is itself responsible for the termination of maternal response.

Summary

Parental behaviour can be divided into three phases: (1) induction; (2) maintenance; and (3) separation or rejection.

The first phase is associated with late pregnancy and parturition. These processes are closely related to the onset of maternal behaviour in most animal species. Endogenous changes presumably under hormonal control appear to trigger the onset of appropriate behaviour towards the young.

The second phase concerns the reasons why the behaviour persists when it has once been established. The presence of young appears to be essential in order that parental response should continue. After the initial processes of the induction of the response the behaviour continues only under the influence of the stimulus provided by the young. Exactly what features of the young have this effect is a matter for further analysis. Certainly a considerable and rapid decline in maternal response takes place if the young are removed.

The final stage is that of separation or rejection. This appears to be brought about largely through the action of the parent. The infant when grown may no longer provide the appropriate stimulus for parental response, or it may provide different kinds of painful or threatening stimuli to the parent which are incompatible with the continued parental response.

Parental behaviour has been studied by examining the response to foster young. In most animals it is uncommon for virgin animals to show parental response to strange young. This usually comes about after parturition of the animal's own first-born young, and persists until the animal no longer responds in a parental way to her own young. Operant methods have been used to study the responses of the female to the young. Virgin primates show little bar-pressing to view infants, but experienced mothers bar-press frequently to view infants. Experience as the result of rearing several generations of young has its effect, principally in the behaviour of the primate female towards her young. Generally, however, the behaviour shown towards the first-born is perfectly adequate, and experience of giving birth and rearing young appears to play a minimal part in the organization of parental response, although other kinds of experience may well do so. The effects of separation of the infant from the mother are described. It is pointed out that there are effects not only upon the development of the infant but also upon the behaviour of the mother.

4   Sexual Behaviour

The task of this chapter is to describe something of the intrinsic mechanisms and the systems of social communication by which animals bring about sexual reproduction.

Sexual behaviour appears to serve several important functions. One of these is to bring the male and female together. Many animals of a variety of species live relatively isolated lives until the reproductive period begins. The problem lies in getting the males and the females together at the appropriate time. This may be achieved by numerous means. Some insect species achieve it by the use of chemical attractants or pheromones, which guide the individuals of one sex to those of the other. Auditory signals may be employed. Birds may attract a mate by the vigour and intensity of their song, insects by stridulation, and mammals through special calls, e.g., those of wolves or orang-utangs. Highly conspicuous structures may be used to attract the mate through visual means. Typical of these are the breeding plumage of many birds and the highly coloured sexual skin of baboons and other monkeys. Frequently striking coloration is accompanied with an elaborate behavioural display designed to reveal the coloration to its full extent. This is so in the behaviour of the peacock who reveals to the hen the magnificence of his tail in his display towards her.

Even in highly social animal communities, distances

may be customarily maintained between males and females. An important function of sexual behaviour in this case is to act as a catalyst, decreasing the distance between potential partners and bringing one within the orbit of the other where this was not the case before. Reproductive behaviour often takes priority over other forms of behaviour which may be incompatible with it. It commonly happens, for example, that adult animals maintain a characteristic distance between one another. Fear, timidity and aggression all seem to contribute to this in their different ways. Sexual behaviour is incompatible with keeping at a distance, and it would seem that the mechanisms of sexual response stand in place of or suppress those which lead animals to remain apart from one another. Aggression creates difficulties in the reproductive situation, and sexual behaviour commonly supplants aggression. The exercise of aggression could bring about a continued failure to mate, although both partners may be in a mating condition. Excessive fear and timidity also are incompatible with sexual response. Timidity must give way to approach if mating is to be achieved.

 The behaviour of one animal has an arousing effect upon another animal. This is easily demonstrated with chicks, when a new member is introduced to a group of previously quiet and sleeply individuals. The previously inactive animals rapidly become alert, giving rise to excited calls. They inspect the new member and then approach to begin the social ceremonies which ultimately determine its place within the group. Just as one chick can excite the activity of others, so it is reasonable to assume that animals of one sex are aroused and excited by animals of the other sex. Sexual arousal may come about through a variety of means. The physical presence of one animal may be sufficient to excite another, but generally speaking any stimulus can have an arousal function, and it may be the case with regard to sexual behaviour that general stimulation from one partner excites sexual response in the other. On the other hand, it may be that there are quite specific signs which have this property and that sexual excitement is brought about by these signs only. These could take the form of particular patterns of behaviour, visual, olfactory or auditory cues, and particularly in the later stages of reproductive behaviour they could take the form of tactile stimuli which would serve this important arousal function.

It is possible that animals may indicate their sexual state to one another through their physical characteristics. The sexual swelling of some monkeys, for example, may act as a sign of their receptive state to other monkeys. Often, however, no such physical signs exist, and in this case behaviour may bear the major burden of communication to the mate or to other animals. In many species the female becomes receptive periodically. At one time she readily accepts the mate but at other times she rejects him. The behaviour she shows during oestrus is different from that in anoestrus. Behaviour clearly indicates her condition to a prospective mate. Behaviour, therefore, appears to act as a signal by which animals can acquire information about the reproductive state of their partner, as well as to indicate by their own behaviour the reproductive condition which they themselves have reached.

The behaviour of one animal may lead to sexual behaviour in another through the process of induction. The courtship behaviour of animals can be seen as a functionally connected chain of events. Signals from the sexual partner are important in causing the animal to move along the chain. At each stage the signals which sexual partners display to one another could act to induce the next appropriate form of behaviour.

The phases of sexual behaviour in both partners inter-link with one another. If they should become out of step, then reproductive behaviour would become difficult. It is essential, therefore, that they are coordinated. This again appears to be achieved by a sexual signalling system which ensures, not only that animals are aroused, and not only that sexual behaviour is induced, but also that this occurs at the right time. It remains as an important problem to be investigated as to how animals are able to phase their activities one with the other and how they bring about the timing of events which this entails.

Sexual influence is not without its influence on nearly every other aspect of the organism's behaviour: status and hierarchy, population dynamics, social groups, aggressive behaviour, ecological systems, and so on. Some aspects of the relationship of sexual behaviour to these factors will be reviewed both in this and in later chapters.

Development of sexual behaviour

Sexual behaviour is not necessarily confined to the behaviour of the adult. Just as Freud pointed to the importance of infantile sexuality in human beings, so many of the patterns of behaviour essential to reproduction make their appearance early in the life of lower animals. This could be seen as a preparation for adulthood. The behaviour is already formed and practised when the time comes to put it into operation. Human infants show erection of the penis from a very early age. Male chimpanzees not only mastur-bate but mount others several years before they become sexually mature (Bingham, 1928). Female monkeys may also show the typical female pattern of crouching and pre-senting the vagina as early as six months after birth

(Carpenter, 1942). Patterns of sexual behaviour make their appearance at different stages during the infancy of the rat. Mounting of other animals appears in sexual play from an early age. Intromission occurs later, and then later still in the adult intromission is accompanied by ejaculation (Rabadeau, 1963). Older rats are less vigorous sexually than younger ones but can, for example, when interrupted by the experimenter during copulation in the case of males, become as vigorous as the young (Larsson, 1963).

Sexual responses

Much of our knowledge about sexual behaviour has been obtained from studies on the laboratory rat. Bermont (1965), for example, analysed the behaviour of the male during intromission by high-speed photography. The male first of all makes a series of thrusts in the region of the vaginal opening. These serve to locate the opening and, when this has been achieved, the male shows a second type of response which consists of a single deep thrust into the vagina followed by immediate withdrawal.

Sexual behaviour in the female is characterized by lordosis or arching of the back. This can be used as a measure of sexual excitability. If females are tested hourly for their reactions to a sexually active male, a different animal on each occasion, then they show a distinct periodicity in the times at which they are sexually receptive which relates to the diurnal light cycle. They are most receptive a few hours before or shortly after the dark phase (Kuehn and Beach, 1963).

Female primates in heat customarily present the genital region as a mating invitation to males. The female usually approaches the male, turns her hindquarters towards him, presents her buttocks and frequently looks at him fixedly over her shoulders. Presentation occurs widely among the old-world monkeys as a preliminary to mating (Ullrich, 1961). Animals which show intense coloration of the sexual skin enhance the effect, and females in heat are generally preferred over other animals (Zuckerman, 1932). Males also show this pattern of presentation to another animal, frequently one which ranks higher in the dominance structure of the group. Sexual mounting may or may not occur under these circumstances.

The effects of experience

Sexual behaviour changes according to the amount of previous experience the animals have. Female cats, for example, become more and more willing to mate as they have more and more experience. At first the female resists the sexual advances of the male. After the first intromission the female no longer resists but takes a more positive part in sexual response. Later, as the female becomes more experienced, mating is initiated more and more quickly after the animals are placed together and the number of intromissions increases (Whalen, 1963). After a series of mating experiences female cats readily accept strange males as partners with which to mate, whereas previously they would have resisted them. Larrson (1959) reports a similar finding with respect to the experienced behaviour of male rats. The number of ejaculations which they show during each hourly period when placed with a receptive female increases as they become experienced in this situation. The animals now ejaculate quicker and need fewer intromissions to reach this state. This effect of experience has been confirmed by other writers, and it appears generally that changes come about in the sexual interaction between mating animals which could relate to the conditioning of the responses to the stimuli provided by the situation, a developing sensitivity to the other animal, or the progressive effects upon the development of skill.

In the experimental study of animal social life one of the most profound influences is that when animals are reared during the early part of their lives in isolation. The effects of early isolation on subsequent sexual behaviour have been closely examined. Strangely enough, the earlier investigations of this factor on rodents failed to reveal any significant effect (Beach, 1942, 1958; King, 1956). Further investigations have revealed differences, however. Zimbardo (1958) showed that male rats reared in isolation, whilst capable of mating, do take longer to mount the female on first introduction, achieve fewer intromissions and copulate less frequently. One reason for this finding could be that animals raised in isolation show severe disorientation. Male rats showed disorientation coupled with an inability to mate when they were introduced to a female in a large area; normal sexual behaviour appeared only if they were confined for mating purposes in a small arena and presumably tactile processes were able to take control more readily under these conditions (Folman and Drori, 1965). Another reason why difficulty in copulating may be encountered by male rats raised in isolation could be that they develop peculiar patterns of motor response which appear to interfere with the normal expression of mating (Gerall *et al.*, 1967). These difficulties do not form a permanent bar to mating, because cohabitation with females for a period of three weeks brings about an effective therapy.

Animals higher on the phyletic scale appear to be far more disturbed by the effects of social isolation than do laboratory rats. It is possible that this is because the higher neural centres become increasingly important in the more highly evolved mammals. Guinea-pigs, for example, fail to show sexual behaviour if they have been reared in isolation (Valenstein, Ross and Young, 1955). Mating is also disturbed in male cats raised in isolation from an early age (Rosenblatt, 1953). In the cat also mating experience appears to play an important part in preserving the mating response in the face of conditions which would otherwise abolish it. Animals castrated after mating experience frequently continue to show sexual behaviour, where as naïve castrated animals fail to do so (Rosenblatt, 1965).

Even small amounts of social experience, as little as 15 minutes a day in a study carried out by Beach (1968) on beagles, was sufficient to bring about normal copulatory behaviour, whereas the behaviour of animals denied this experience was impaired. The socially deprived animals attempted to mate as frequently as the other animals, but the orientation of mating was frequently wrong, the male attempting to mount the female's head, side or flank. This result suggests that social isolation has no apparent effect upon sexual drive, but that the social communication and patterns of motor coordination between one animal and another are disturbed. It has been reported that sexual disturbances make their appearance in primates as the result of social isolation (Harlow, 1965; Mason, 1960; Hansen, 1962; Harlow and Harlow, 1962). Meier (1965) has failed to confirm this result, reporting that animals reared in isolation from the first day of life were able to mate in a perfectly adequate manner. Meier points out in this study that a possibility existed for visual and auditory stimulation, although not tactual between the isolated juvenile animals. It may be that communication of this kind is important in bringing about subsequent mating. It is certainly a result which needs further investigation. The results of studies on social deprivation as a whole, and their effect upon later sexual behaviour, suggests either that patterns of behaviour develop or are learned during this period of isolation which are incompatible with sexual social responses, or that an important opportunity has been denied the isolated animals to learn the patterns of communication, as well as those of motor response by which adequate mating is brought about.

Learned aspects of sexual response

The opportunity to carry out some aspects of sexual response can in many instances provide a powerful source of reinforcement for learning. Classical conditioning, for example, can be carried out using the sexual object as the conditioned stimulus. Sokolova (1940) introduced a castrate ram to other rams. A bell was sounded each time they mounted the castrate in a sexual manner. At the fifty-fourth day of training the bell was sounded in the absence of the castrate ram, the sexual object, and now the active ram exhibited erection of the penis as an anticipatory response.

Sheffield, Wulff and Barker (1951) showed that instrumental response from male rats, such as running an alleyway to a female in heat, can be stamped in by aspects of the copulatory pattern. If, however, male rats are allowed to run down an alleyway to a female and then only allowed to mount, but not to achieve intromission, they run progressively slower on future occasions (Whalen, 1961).

Female rats, on the other hand, run an alleyway to reach a male, but it makes no significant difference to the change in running time if the male is one which will mate with them or not. Mating appears not to reinforce instrumental behaviour in the female rat in the same way as it does with the male (Bolles, Rapp and White, 1968). Sexual behaviour, particularly in the male, is capable of motivating different aspects of instrumental behaviour. Aversive learning can also exert continuing effects on performance which relate to behavioural pathology in animals. If newly weaned male rats have an electric shock administered to them each time they attempt to approach an adult female, then when the animals reach maturity they continue to avoid the females. Heterosexual activity is thus disturbed by the aspects of early experience. One interesting question which arose from this work concerned the nature of the sexual behaviour which these animals show if their usual approach to the female is blocked. Hayward (1957) reported that no substitute homosexual activity appeared.

The effects of interruption on sexual behaviour

Sexual behaviour in the male rat takes the form of a series of pelvic thrusts or intromissions which increase in frequency at first during copulation, but then decrease as ejaculation is approached (Dewsbury, 1967). One important way of examining sexual motivation is to interrupt animals during copulation between one intromission and the next with the aim of seeing how long it takes the animal to copulate after interruption. Brief interruptions act to speed copulation (Larsson, 1959), whereas longer interruptions produce no further effect (Beach and Whalen, 1959).

It is possible also to see what aspects of sexual behaviour are rewarding to the animal in terms of the degree to which further sexual behaviour is promoted. If male rats are allowed to mount female rats, but are removed before they can penetrate the female, then it is found that the males take longer and longer on subsequent occasions to initiate mating. This behaviour is no longer sufficiently rewarding. If, however, the males are allowed to mount the female and to have vaginal intromission before the experimenter interrupts their behaviour, then some reward seems to have been provided because the animals subsequently persist in this behaviour even though they may not be allowed to approach ejaculation (Bermont, 1964). Evidence of this kind points to the view that ejaculation is not the only source of reward to the male rat in the copulatory sequence. The latter are some of the results which have been obtained, using the technique of interrupting sexual behaviour. It is a powerful technique which can provide quite a lot of evidence about sexual motive states.

Exhaustion and recovery

Animals may not be able to maintain continuously high levels of sexual performance. They become exhausted frequently and need to recover before initiating mating again. A cock, for example, released into a cage of hens shows intense courtship activity during the first three minutes. After that a decline sets in and courtship activity becomes less and less frequent. When successive males are presented to them the females also become less and less sexually active. They usually show sexual crouching in response to the cock, but this declines with the successive presentation of each male.

Soulairac (1952) has clearly shown that after rats have copulated, there is a period during which they rest and do not copulate again. There is also an increase in this so-called refractory period with successive ejaculations. Hamsters also suffer exhaustion after sexual activity and only fully recover from complete exhaustion after a period of five days (Beach and Rabadeau, 1959).

There are, however, examples where this period of exhaustion can be curtailed. Animals may be previously exhausted by one sexual partner. If, however, they are now introduced to a fresh partner, they may begin mating again and recover much of their vigour. Presumably the original partner no longer has the capacity to arouse sexual response, whereas a fresh partner has different stimulating properties. Male cats having mated to exhaustion show a recovery of sexual function when placed with fresh females (Whalen, 1963). The same author suggests that sexual arousal may be analysed into two components, one which is specific to each animal, to each male and to each female, as well as one which is a function of the interaction between the partners.

Sexual preference

One of the most successful techniques has been one in which the animal is able to control its environment and thus demonstrate its own preference for animals with which to mate, or its own choice of particular rates for sexual responding. A number of investigators have trained rats by presenting them with a lever which when pressed will release an animal of the opposite sex into the cage. Schwartz (1956) trained male rats to press a lever to secure sexually receptive females. If the males had been somewhat sexually deprived prior to the experiment, then they pressed the lever frequently. As they became satiated, however, the rate of lever pressing fell off dramatically. If the roles are reversed and a female is allowed to press a lever which will admit a sexually active male to her cage, then lever pressing varies as a direct function of the stage at which the female has reached in her oestrous cycle.

If the female rat is allowed to regulate her contact with the male by the provision of an escape compartment which she can enter without being followed by the male, then it is found that she chooses to avoid the male immediately after copulation has taken place, and in any one test session she spends longer in isolation at the end than at the beginning of the session. Pierce and Nuttall (1961) suggest that there may be an aversive as well as an approach component in the sexual behaviour of the female rat. The female also, when provided with a bar to press to secure a sexually active male, delays pressing for a considerable time after copulation, but presses very rapidly if copulation has been interrupted (Bermont 1961).

The time spent by male dogs near a cage containing females varied considerably as to whether they were in oestrous or not. Also if male dogs were tethered in an enclosure and the females allowed to roam freely amongst them, it was found that some males receive more female attention than others. Preferred males were sought most frequently, allowed to mount earlier, and were rarely barked at or bitten whilst investigating the female (Le Boeuf, 1967). Operant methods and those using instrumental response have been used successfully to highlight important aspects of sexual behaviour.

Sensory factors in the control of sexual behaviour

Visual effects have an important influence on a number of aspects of sexual behaviour. One way of examining these effects is to construct models of various kinds which act to release or induce sexual behaviour in the animal to which they are presented. By varying the characteristics of the model it is possible to pinpoint those characteristics of it which are important in bringing about sexual response and those which are not. Lack (1939), for example, found that stuffed robins without the red-breast coloration are frequently courted by male robins, whereas those with the red-breast coloration are attacked.

Models of domestic hens possessing some features but not others were constructed by Carbaugh, Schein and Hale (1962). The cock behaved in a sexual fashion to these models. It was possible by altering aspects of the models also in this case to assess the effects of the various parts of the model in releasing sexual response, by removing parts of it or adding parts elsewhere. When the tail was removed, sexual response of the cocks remained at very much the same level. The tail therefore appeared to play little part in releasing sexual response. The body is necessary for the release of sexual patterns, and the head appears as a signal to control proper orientation towards the model, because in the absence of the head sexual behaviour could be directed to either end of the model.

The sense of smell is of importance in arousing sexual response in a number of mammalian species. Rats, for example, show a distinct preference for the odour of receptive as opposed to non-receptive females (Carr and Caul, 1962). Male dogs also appear to be highly stimulated by the odour from the urine of a bitch in heat (Beach and Gilmore, 1949). Males spend considerably longer sniffing the urine from receptive females. The importance of olfactory communication in mice has been emphasized by Bruce and Parrot (1960). Bruce (1961) showed that pregnancy in female mice is frequently blocked if a strange male is placed with the female after mating has occurred with another male. It is not necessary that the strange male is actually present in the cage with the female. If materials with which the male has previously been in contact smelling of the strange male are exposed to the female then the same effect is produced (Bruce and Parrot, 1961). Therefore the smell of strange males and the smell alone is sufficient as a signal to block pregnancy.

Sexual behaviour is brought about in a number of insect species by a variety of ectohormones or pheromones, chemical substances which act as sexual attractants. Butler (1967) has reviewed the evidence that substances of this kind operate in a variety of species of moth, butterfly, fly, beetle and cockroach. In the case of the silkworm, for example, the males react to the extract from the female silkworm with copulatory behaviour and a high level of sexual excitement. Michael and Keverne (1968) suggested on the basis of their investigations that some similar substance may play a part in the reproductive behaviour of primates.

One technique for observing the contribution of the sensory systems to sexual behaviour lies in the use of surgical interference to render animals blind, deaf or anosmic. Beach (1942) showed that a rat deprived of its sense of smell by transection of the olfactory nerves will readily mate if it is allowed contact with a receptive female. Partial deafening does not prevent sexual behaviour in the rat (Beach, 1951), nor does clipping the vibrissae or desensitizing the skin of the snout and the lips (Beach, 1942). Brooks (1937) showed also that surgical interference with the visual system in rabbits does not prevent the female from receiving the male.

In many animals sexual behaviour in sensorily deprived specimens can be maintained if they are placed in physical contact with one another. Tactile and pressure stimuli would seem to play a part. However, the evidence is equivocal on this point because female rats lacking genital sensations may still show some sexual behaviour. Beach (1945) found that a particular female with congenital absence of the uterus and the vagina was nonetheless receptive to the male. Kaufman (1953) removed the ovaries, uterus and vagina in female rats. These animals under special conditions of hormone treatment showed receptive behaviour when paired with normal male rats. Females which have been anaesthetized in the vaginal and perineal regions also show receptive behaviour, but there appears to be an effect upon the performance of the male because ejaculation is delayed (Bermont and Westbrook, 1966). When the penis of male rats is anaesthetized, they continue to mount with relatively high frequency but experience difficulty in intromission. If the penis in male rabbits is extirpated, then animals show every sign of sexual excitement and will make repeated attempts to mount the female. Evidence of this kind suggests that whilst the physical stimulus of contact may be important, nonetheless sexual activity is induced in its absence, and the main sources of sexual drive appear to be related to it in only a secondary fashion.

The evidence from this analysis of sensory factors with regard to sexual performance in higher mammals suggests that performance is more difficult in the absence of appropriate physical contact or in the absence of genital stimulation, but nonetheless sexual behaviour still occurs, although of a somewhat ineffectual and disoriented kind.

Physiological factors in sexual behaviour

Many investigators have concerned themselves with the relationship of sexual behaviour to various aspects of nervous function. Several aspects of male sexual behaviour in animals appear to be controlled at the level of the spinal cord. In dogs, for example, the ejaculatory reaction and the copulatory lock appear to be mediated at this level (Hart, 1967). Female cats also may show such typically sexual patterns as treading and moving the tail to one side as in copulation, even after transection of the spinal cord. In many female animals mating occurs although the whole of the cerebral cortex has been removed. The integrity of higher brain centres appears to be more important when the sexual behaviour of the male is considered (Beach, 1940, 1941). The mid-brain region appears to be particularly closely associated with the control and arousal of sexual behaviour. Lesions in this area (Lisk, 1966) and electrical stimulation of the anterior dorsolateral hypothalamus produced a marked increase in the sexual capacity of some rats (Vaughan and Fisher, 1962). During electrical stimulation the rats would press a bar to open a door to obtain access to females. Even after the males had ejaculated they continued to open the door to obtain access to females and to display sexual activity whilst the electrode continued to stimulate their brain; as soon as the stimulus terminated, so did the males' sexual activity. Studies of this kind point to the association of the so-called 'pleasure' area of the hypothalamus with sexual responsiveness. It is possible that hypothalamic mechanisms are responsible in some way for the control and maintenance of sexual drive.

The reproductive hormones have shown themselves also to be of considerable importance in relation to sexual motivation. One finding relates to the fact that hormones may have an influence early in the life of the organism. Androgens administered to guinea-pigs prenatally by injection through the mother's perineal region not only act to masculinize the external genitalia, but also masculinize the behaviour of female guinea-pigs such that in many ways they now resemble the male (Goy, Bridson and Young, 1964; Gerall, 1966).

The gonadal hormones play an important part in maintaining sexual behaviour, and this fact is illustrated in the changes which come about following castration. Post-operative changes can often be reversed by administering the appropriate hormone. An animal incapable of mating after castration may become capable after, e.g., the administration of androgens (Beach and Holtz-Tucker, 1949).

The oestrogens or female hormones appear to play a less powerful part in the control of sexual response than do the androgens or male hormones. Male rats treated with oestrogens show little in the way of female behaviour (Feder and Whalen, 1965). Similar results are also obtained with castrated rats. Feminization is produced, these authors argue, not through the presence of oestrogen but by the lack of neonatal androgen. In female rats, however, there appears to be little doubt that sexual behaviour relates closely to the oestrogen level (Freedman and Rosvold, 1962). Oestrogens appear to lose their effectiveness in changing sexual response shortly after fertilization in animals which have become pregnant or in those which are pseudopregnant. Female sex behaviour would appear to be inhibited under conditions such as these where other progestational hormones are at a high level (Powers and Zucker, 1967). Hormones then play a crucial role in the control of many aspects of sexual behaviour, both in determining sex typical patterns from an early age and in acting as motivational spur in later life.

Summary

In this chapter we have been concerned with a dual process, on the one hand with the bodily mechanisms of sexual response, and on the other with the nature of the stimuli which call forth sexual behaviour. We have examined the reflex systems, the lower centres of the spinal cord, as well as those of the mid-brain region. We have examined the system of somatic sexual hormones as well as aspects of sexual drive. In doing this we have attempted to provide some explanation of the basic physiological processes underlying sexual response, but if we were to leave the question there, we would have provided at best only a partial explanation of sexual behaviour, because it must be remembered that mating is a form of social response, it is directed towards other individuals, it involves an interaction and a coordination of response with them. It is essential, therefore, to consider sexual behaviour from this point of view as well as to discuss its structural facets.

Sexual behaviour can be regarded as a series of interrelated events leading through the process of courtship and preliminary forms of behaviour to ejaculation or orgasm as the end response. The events in the chain could be regarded as the switching in of different physiological mechanisms. Even so, many external sensory factors enter into the system, integrating themselves within the chain of response. Sensory factors act as the trigger to many of the patterns of the behaviour, causing movement from the first links to the subsequent ones of the chain. At each stage in this chain, sensory factors enter in to modify the type of behaviour produced, and the evidence points to a progressive switching or addition of stimulation from different modalities as sexual behaviour moves on towards its climax. Sensory information could both act as a specific signal to evoke fresh responses or it could change the basic state of motivation of the animal, as well as providing a means to control and guide the behaviour forms which are dependent on it.

The signals which act as the spur to sexual response are largely those associated with the sexual pattern. Each animal is not only involved in a sensory link with the other, but in a communication using quite specific signs that act as a language of courtship which ultimately brings mating about.

Lorenz (1963) has discussed aggression as the attack by one animal on a member of its own species. The behaviour shown by an animal when it attacks a member of another species may, however, be similar in all essential respects. It would be arbitrary to restrict the definition of aggression to the behaviour shown in the former but not the latter case. Aggression should perhaps be defined in terms of the behaviour itself and not the object to which it is directed. Aggression can then be seen as a form of behaviour having the characteristics of threat, hostility and attack.

Attack may appear as a corollary of competition. It may appear when animals find themselves in competition for food. It may occur in other competitive situations, such as those for sexual possession, an attempt to gain territory or simply to acquire status or exert authority over another animal. Attack may also occur in situations not directly related to these particular ends. Attack both of a directed and controlled or uncontrolled kind can occur in response to the stimulus of pain. Presumably attack in this case occurs normally as part of the defence against other attack under circumstances in which one animal has inflicted some form of minor injury upon another. Attack also occurs, however, towards any convenient object when animals are exposed to painful stimuli for experimental reasons. Attack may occur as a response to or as an attempt to relieve frustrating circumstances. Here, again, attack appears to relate to the usual competitive encounter where the dominance or success of one animal may be achieved only through frustration of another animal. Attack not only against other animals, but also against convenient objects is to be observed in situations arranged by experimental processes to provide frustration through the withholding of food reward or by some other means.

There are many ways by which animals direct hostility to one another. Aggression need not be expressed only by overt attack. Frequently a series of hostile signals are shown which stand in place of attack and may act as a warning that attack is imminent. These are threat displays which can be sufficiently subtle to go undetected by all except the most sensitive observer, or they may be large-scale responses which are unmistakable to all. Threat displays signal aggression, and they may be sufficient in themselves to change the relationships which one animal holds to another. If they are ineffective in this respect, then they may form a prelude to more violent forms of aggression, each acting as a signal nonetheless of the hostile states. Threat displays appear to form an important means to signal aggression and in many cases to reduce hostility, because the exchange of signals can often stand in place of physical attack.

Paradoxical though it may seem, the evidence points to the fact that aggression has the well-being of the individual as one of its essential functions. It is used to protect the family and social group as well as the individual. It is a method for establishing and preserving the supply lines for each family. Aggression, whilst it may result in harm to animals when fighting breaks out, does serve to protect individuals when threat takes the place of physical combat.

Patterns of hostility

Aggression has been the subject of extensive investigation over the years, both because of its prominence in the behaviour of animals and because of its importance in relation to the behaviour of man. A few examples devoted to the study of aggression are presented here to illustrate the nature of this work.

Aggression is widespread amongst insect species and in some animals it appears to relate to their colonial habit. Members of different species and members of different colonies may be extremely hostile to one another.

An alien ant entering the nest may be threatened and subjected to prolonged seizure and dragging. Attack in some cases can be elicited by the presence of formic acid (Vowles, 1952), and it is widely believed that chemical stimulation plays a part in the regulation of aggressive behaviour, and that it is through the chemical senses that the animals appear to be able to distinguish strange animals from those of their own colony. Licking of alien animals may occur, as well as licking of returning nest mates, and it may be that this is the way in which discrimination is established (Wallis, 1962).

Allee (1931) also points to the possible importance of chemical factors in determining the aggressive reaction to strange ants of another colony. He found that ants imported into an alien colony when young will not attack the members of that colony when subsequently placed in physical contact with them, and this suggests that the early chemical relationships with the colony are important in determining whether attack will occur.

Not all aggression, however, relates exclusively to the chemical senses. Tinbergen, in a famous classic, *Social*

Behaviour in Animals, describes how visual signals are important in the production of aggressive behaviour in fish. Tinbergen describes the territory formation of stickle-backs. Aggression is related to this because in the defence of territory the male shows the characteristic threat posture. This is a vertical attitude in which the head points downwards, and the flank or underside is directed towards the opponent, whilst one or both of the vertical spines are elevated. Tinbergen, in a series of well-known experiments, showed that the red coloration of the belly was an important stimulus to elicit fighting. He constructed a series of models. Some models were life-like, others barely resembled sticklebacks at all. When the belly region of the model was coloured red, then the model was attacked by a male stickleback. Models not showing this coloration were hardly ever attacked. Tinbergen argued that the red coloration was a releasing stimulus for fighting behaviour in the male.

Many other fish besides the stickleback show interesting patterns of behaviour in aggression towards others, and these also may be released by specific stimuli. The Siamese fighting fish (*Betta splendens*), for example, shows a most colourful display as part of its hostile behaviour. This display occurs in response to the stimulus provided by other animals, but it can also be called out by the fish's own image when seen in a mirror. When male Siamese fighting fish are paired, one fish approaches the other rapidly and directly. The fins spread and the gill covers become erect. The fish turns sideways to its opponent, ensuring that the maximum area of effective display is exposed to it. Both wild males and females darken in colour in aggressive encounters until the body is a velvety red-black and the head and dorsal parts of the body are nearly jet-black (Simpson,

1968). The encounter is often prolonged in males and serious injury may result, but in females the encounter ends when one member of the pair stops displaying. The loser then moves rapidly away, chased by the other animal (Braddock and Braddock, 1955, 1958). The loser usually moves away rapidly on other later occasions and the relationship becomes crystallized into a status order.

Visual signals are not the only ones which are used to express hostility and aggression. Moynihan (1959, 1962), for example, has described the hostile calls of a number of gull species, analysing them in terms of their motivational properties particularly in relation to the balance between the 'drives' of attack and escape. Most gulls give rise to calls which express different degrees of hostility, and a typical example of this is the 'pumping call'. It consists of a series of harsh rasping notes uttered in a sequence when one bird is attacking another or when the birds are in dispute. In these birds hostile calls accompany certain aggressive bodily postures. Hostile calls are associated with a downward hunched posture or with the low-oblique postures of adult birds. Various patterns of hostile behaviour appear to develop whilst the young are in the nest, and by the time the animal first begins to fly it is equipped with an extensive repertoire of various aggressive calls and displays.

Aggression and hostility can be expressed through a variety of different sensory modes. Most animals show typical patterns of aggression; however, these may not always take the form of attack. Frequently hostile displays are sufficient. Species appear to be equipped with characteristic displays which allow rapid communication between one animal and another of the intensity of aggression and of the probability of future attack.

Aggressive behaviour and brain function

The display of aggressive behaviour bears a direct relation to interference with closely prescribed regions of the brain. In studies of surgical interference and electrical stimulation it has been reported that complete patterns of aggressive behaviour appear in otherwise tranquil animals, when these techniques are applied to the hypothalamic areas. Goltz (1892) found in dogs, and Bard (1928, 1934) in cats, that removal of large parts of the fore-brain appears to release intense emotional response from the control which had formerly existed. Decorticate animals show emotional hyperactivity, in which the slightest sensory stimulus leads the animal to become exceedingly aggressive. Studies of this kind suggested that the hypothalamic mechanisms are related to the expression of aggression and, because the rage shown by decorticate animals appeared to have little direction or to relate to conditions normally giving rise to aggression, it became known as 'sham-rage'. Cats with large hypothalamic lesions show the typical patterns of growling, crying, spitting and erection of the hairs, particularly those along the back and on the neck (Kelly, Beaton and Magoun, 1946). It is now obvious that not all of the hypothalamus is responsible for this effect, and attempts have been made to state more precisely those areas which relate to the expression of aggressive behaviour (Anand and Brobeck, 1951; Karli and Vergnes, 1964).

The presence of non-directed rage as the result of the release of hypothalamic centres from higher control again suggests that the object of aggression should concern us less than the patterns of behaviour themselves. Certainly, in the case of decorticate animals, minor stimuli are sufficient to evoke rage in the absence of the normal provoking situation. It is possible that in this case the complete sequence of aggression is laid down as a behaviour form under the control of the hypothalamic system, but that removal of the cortex also removes the usual guidance mechanisms. The behaviour is preserved but is blind and uncontrolled in its expression.

It has been reported that other brain regions are associated with the control of aggression (Kennard, 1955; Pribram and Bagshaw, 1953). The hypothalamus and related area would appear, however, to play the major role. There are many difficulties associated with surgical studies which must be borne in mind when interpreting the results of this kind of investigation. A structure playing an integral part in the control of a particular pattern of behaviour in one species may not, because of differences in brain organization, play that same part in another species. Lesions may not only disrupt the control mechanisms of the behaviour under study but also other systems, for example those associated with sensory or motor proficiency in addition to those concerned with the coordination and control of behaviour patterns. Small differences also in the size of the lesion may result in profound differences in behaviour. For these reasons the results obtained through the use of the technique of implanted electrodes are often to be preferred over those obtained from ablation studies.

The use of implanted electrodes is a technique which has been widely used to study aggression as well as other forms of behaviour in animals. Von Holst and Von Saint Paul (1963) report investigations in which electrodes were implanted in the brain of cockerels. Whole sequences of behaviour of a coordinated kind were elicited by stimulation. Aggression had a coordinated character and was directed in this instance towards a stuffed polecat provided by the experimenters. The absence of the polecat did not render the animals any the less aggressive, and after hypothalamic stimulation they directed their aggression towards the face of the experimenter as a substitute. Hess (1928, 1954) was one of the first investigators to describe the effects of hypothalamic stimulation in cats. When the cat is stimulated through deep-lying electrodes in its hypothalamus, it shows, with the exception of arching of the back, the pattern of response which it would normally show in aggression against a dog: the pupils widen, the ears lie flat against the side of the head and the tail becomes bushy. If stimulation is maintained or particularly intense stimulation is used, the animal attacks objects around it and will attack a person standing nearby.

It is possible using the technique of electrical stimulation to evoke attack without rage, as well as rage without attack, and thus separate these two factors from one another.

Wassman and Flynn (1962), by stimulating one area, induced cats to attack mice which they had not previously attacked and, by stimulating another, they induced these animals to show considerable rage, as judged by their behaviour; yet under these circumstances the mice remained unharmed. There is some evidence, however, to indicate that animals under the influence of hypothalamic stimulation are not without some powers of discrimination with regard to the objects which they attack. Cats, as the result of hypothalamic stimulation, attacked not only rats which had been anaesthetized, but also stuffed rats, far more frequently than inanimate objects such as a toy dog or a block of foam rubber (Levison and Flynn, 1965). There is some evidence to suggest also that vision is relatively unimportant in determining whether attack will occur in the laboratory rat as the result of hypothalamic stimulation (Levison and Flynn, 1965; McDonnell and Flynn, 1966). If the animals are confined in a small space, then one animal attacks another and removal of the animal's vision has no effect upon this. Olfaction appears not to play any great part either, but denervation of the snout reduces the severity of the attack.

The predominance of attack under hypothalamic stimulation over the other behaviour mechanisms is illustrated by the fact that if a cat were eating, for example, and then hypothalamic stimulation were administered, the animal would turn immediately and attack a nearby rat (Roberts and Kiess, 1964). The opportunity to attack under hypothalamic stimulation is also reinforcing, because cats readily learned to run a maze to obtain a rat to attack again under hypothalamic stimulation, but if this were not provided the running habit rapidly deteriorated.

This evidence, particularly that from electrode implantation, suggests that the control and motivation of aggressive behaviour is associated with the hypothalamus. In many cases the resulting aggression is not fully directed or may be directed to any convenient object. This suggests that coordinated mechanisms for aggression exist within the brain and that the effect of the stimuli is to call these forth. The fact that sequences of coordinated behaviour may be evoked, which can often last longer than the stimulus itself, again suggests that the hypothalamic stimulation provides a trigger to the system, implementing preformed patterns of response. It would seem that the animal brings to social situations sequences of aggressive behaviour which are already elaborated in association with the hypothalamus. In the usual course of events, these lie dormant, but they can be called out when the occasion arises.

Hormones and aggressive behaviour

Compared with the central hypothalamic neural mechanisms, hormone systems appear to play a rather more restricted part in the control of aggressive behaviour.

The major influence which they bring to bear, as far as is known, is through the association of the presence of aggression with male sex hormones. The effect of castration in the male in a variety of mouse species (Tollman and King, 1956;; Bevan, Dawes and Levy, 1960) is severely to reduce aggression and to change the animal so that it becomes more quiet and passive. The effect of replacing androgen is to restore the animal's status in aggressive encounters with other mice to the level attained before castration.

Castration has similar effects upon rats (Karli, 1958), causing them to assume subsequently a much lower position in the social hierarchy as determined by aggressive encounters. Whilst castration affects aggression relative to members of the animal's own species, the experiment by Karli showed that neither castration nor the administration of testosterone had any effect on the propensity of rats to kill mice placed in the same cage.

This appears to support Lorenz's view and suggests that the hormone effect may relate only to aggression displayed between animals of the same species and possibly to that shown primarily in sexual competition. The effect of testosterone in leading the animal to behave in a characteristically male fashion, in which aggression is an important feature, is most strongly felt early in the life of the organism (Edwards, 1968).

If testosterone is administered early in life, then the adult behaves in a more aggressive fashion. If testosterone

is administered at birth to female rats, 90 per cent of the
animals fight subsequently in a paired encounter with
another animal in a strange cage. Hormones appear to
exert their major influence in this respect early in the life
of the animal. Hormones of this kind also seem to be largely
responsible for the frequently observed differences in
aggressive behaviour between male and female animals.
Presumably these characteristic differences are laid down
by the organizing influence of the hormone from an early
age.

Aggression and sensory stimulation

We have seen previously that olfactory, visual and auditory
signals may be used to communicate hostility, and that
aggression can occur as a response to apparently specific
signals which have the effect of releasing or activating it.
The investigations of hypothalamic mechanisms suggest
the presence of prepotent behaviour patterns which become triggered by specific patterns of sensory stimulation.
The question can now be asked as to what are the effective
stimuli in bringing about aggressive response. There are
several well-known examples of the release of aggression
by stimuli possessing quite definite and specific characteristics. Tinbergen showed this in his classic example of the
release of fighting response by the red coloration of the
belly of the stickleback. Lack (1943) also showed that fighting is elicited by the red coloration of the breast of the
male robin. Stuffed male robins attached to a post are
attacked by territory owners in the breeding season.

It is possible that stimuli of this kind simply attract the
attention of the animal and in this way act as a focus for

aggressive behaviour which has become aroused through other means; in this context the interference produced by the experimenter cannot always be ruled out. It is also possible that the sensory systems themselves filter out some areas of stimulation, leaving a narrow and restricted region as the sensitive area. Stimulation within this region could either act as a key to unlock the pattern of behaviour or it could alternatively arouse it from a dormant state, but in either case the behaviour would be related to the stimuli by virtue of the limitations imposed by the sensory systems.

Wolff (1965) began an investigation of the contribution of the sensory systems to the release of aggressive behaviour of rats, by progressively interfering with different sensory modalities. Rats can be reliably induced to fight if they are placed in the same cage and one or both are administered an electric shock. Fighting is considerably reduced in animals whose vision is removed by temporary occlusion or by permanent surgical interference, and aggression is considerably reduced in animals whose vibrissae have been removed.

Experiments of this kind not only point to the importance of the visual and tactile sensory systems, but they also illustrate the contribution of one of the most important systems of all in the production of aggression – that is, the system by which the animal experiences pain. Painful stimuli reliably elicit aggression and it may be that what we have been considering previously can be explained by the fact that painful stimulation produces conditioning of the most rapid kind. Aggression is therefore likely to occur frequently as a conditional response to any stimuli which have been associated in the past with pain. Scott and Frederickson (1951), for example, showed that painful stimulation during infancy is an important factor in the early appearance of aggression in the grey mouse. Presumably also painful stimulation occurs for the large part as the result of the actions, or at least in the presence, of other animals. They could in this case become conditioned stimuli for the presence of pain. Different behaviour patterns shown by them become conditioned to a greater or less degree, and become the signals for aggression on the part of the original animal.

The fact that aggression occurs in response to pain has been established in many investigations. Tedeschi *et al.* (1959) were able to elicit aggression in mice housed in the same cage if they were exposed to mild footshock. Ulrich (1961) also reported this finding for rats. Hutchinson, Ulrich and Azrin (1965) report that rats reared in isolation show considerably less aggression than socially reared animals, consequent upon shock. This finding again suggests that early conditioning in response to pain is important in the development of aggression between one animal and another. Squirrel monkeys show aggression towards one another under the influence of pain (Azrin, Hutchinson and Hake, 1963), and attack can be elicited from these animals by exposing them to tail pinches (Azrin, Hake and Hutchinson, 1963). Attack appears to be directed towards any convenient object. If a coloured cloth ball was suspended in the cage, then attack was directed towards it. A stuffed doll placed in the cage was attacked. Rats, mice and other monkeys were also attacked. The probability of attack is a direct function of shock intensity over the range tested.

These results not only point to the importance of pain as being directly responsible for the production of aggression but they also suggest that pain is associated with early conditioning, such that signals in the behaviour of one animal release conditioned aggression in the other.

Learning and the modification of aggressive behaviour

The contribution of learning to the study of aggression is something which needs further appraisal. Stimuli which originally had a minimal or neutral effect upon the behaviour of an animal may now, after learning has taken place, be capable of calling out full-scale aggression. It is necessary, not only to establish how this comes about, but also to assess how far the behaviour of animals is governed under normal circumstances by the transfer of aggression as a behavioural process to new or additional kinds of signals, which act in place of the previous ones.

Classical conditioning of aggressive behaviour can be established by pairing tones or other stimuli with those situations giving rise to aggressive response. Aggression then occurs after training to the new stimulus. Rats can easily be conditioned by this method. When pairs of rats are placed in a cage, aggression is evoked between them if an electric shock is given to them. If a tone or a buzzer accompanies the electric shock, then after a few trials aggression may occur not only in response to the shock but in response to the buzzer alone (Creer *et al.*, 1966; Vernon and Ulrich, 1966).

Fighting appears to be not only relatively easy to condition but also to be persistent when once established. Willis *et al.* (1966) by the use of operant techniques trained one pigeon to peck at another pigeon in order to obtain food reward. Continued aggression was observed in animals trained in this manner. A bird so trained was observed to fight a variety of opponents at a period of 11 months after the original training. The hen dominated all opponents except one highly aggressive cock which showed a pre-mating behavioural response in her presence. The use of this technique again illustrates the use to which operant techniques can be put in the study of animal behaviour. In this case the technique creates a means of analysis of fighting behaviour.

The study of pigeons in operant situations also illustrates another highly important condition responsible in giving rise to aggression – that is, frustration. When pigeons have been frustrated, they frequently turn and attack other pigeons placed in the same cage with them. After pigeons had been conditioned to peck a key to obtain food reward, they were frustrated because the food reward was withheld. Under these circumstances they turned to other pigeons and attacked them although the other pigeons were in no way responsible for the lack of food (Azrin, Hutchinson and Hake, 1966).

Amsel (1962) found that a rat will show an increase in running speed in a second runway joined to the first, if it has failed on that trial to find food reward in the expected place at the end of the first runway. Amsel has described this as a frustrative non-reward situation. Animals also show aggression in this situation, if they have been frustrated by the failure to find food reward in the first goal box (Gallup, 1965).

Another effect associated with frustration occurs when animals have been previously trained to carry out a particular task for food reward and then the training is extinguished by withholding the reward. Immediately the extinction phase has begun, rats show a temporary increase in pressing the lever which previously provided them with food, but at the same time they also show a greater tendency to turn and attack other animals (Thompson and Bloom, 1966). Both tendencies could be regarded as part of the aggressive response to the conditions of frustration which have been imposed.

These are examples of the effect of frustration in experimental conditions. Frustration also occurs, however, under natural circumstances. When animals find themselves in competition, the success of one represents a frustration to another. Aggression would be expected on many occasions as a corollary of this, and certainly frustration cannot be overlooked as one of the important reasons for aggressive response.

Closely related to the problem of the learning of aggression and the transfer of aggressive response from one stimulus to another, is the problem whether aggression may be abolished or suppressed by particular forms of training. Several attempts have been made to modify social aggression through appropriate conditioning techniques. Aggressive behaviour can be rapidly and easily extinguished in cockerels if each time one particular animal attacks one of its flockmates it is given an electric shock (Radlow, Hale and Smith, 1958). Dominance relationships in the flock were disturbed by this procedure, and animals in the flock which had been previously dominated now themselves dominated the cock, which had been trained not to be aggressive.

One of the responses which has been studied in relation to the suppression of aggression is that of mouse-killing by rats. Some rats consistently kill mice which have been placed in their cage. If painful electric shock is administered to these rats when they begin to kill mice, then this response can be suppressed (Myer and Baenninger, 1966). Shock administered randomly in such a manner that it was not related to killing behaviour or was not contingent upon it, had little effect in suppressing the response.

Baenninger (1967) looked further at the problem of mouse-killing by rats. The painful electric shock was administered consequent upon killing, and a buzzer was sounded in conjunction with the shock. The sound of the buzzer alone on subsequent occasions was sufficient to inhibit aggression. Aggression in this case has been brought under control by the training technique, and a formerly neutral stimulus (the buzzer) now acts to prevent the rat killing the mouse. Myer (1967) points out, however, that some acts of aggression are more difficult to suppress than others. Rats, for example, with extensive experience of killing mice give up this pattern of behaviour only slowly. Habits which through much previous experience have become ingrained are more difficult and take a longer time to suppress, but even these ultimately yield to the effects of aversive training and the new relationships to punishment which brings this about.

These investigations have shown that the suppression of aggression can occur in a situation controlled by the experimenter. The question has now been asked whether suppression of aggression can occur in situations controlled by other animals. In other words, will animals work in a social operant situation to suppress aggression in other animals? Delgado (1963) implanted electrodes in the caudate nucleus of a large male monkey. Electrical stimulation of this area of the brain inhibits aggression. The monkey used in this study was the chief or dominant male monkey of a colony. This monkey had been conditioned to be aggressive, and this behaviour produced escape and fear responses from all the other animals in the colony. Using a telemetric system a lever was provided which could be pressed by a subordinate monkey. The effect of this was to stimulate electrically the caudate nucleus in the brain of the boss monkey, causing his aggression to give place to more peaceable behaviour. The subordinate monkey repeatedly pressed the lever to control the dominant monkey's aggression. This study shows that one animal will work to effect a change in another animal's behaviour. This result suggests that fear of the behaviour of one animal may motivate the performance of another animal.

Work of this kind has widespread application, not only in understanding the mechanisms of animal behaviour, but also in illustrating possible ways in which aggression might be diminished and controlled in man. This relates to the clinical and educational problems of the individual as well as the larger political questions of conflict and war. Further than this, the idea that the aggression of animals may be controlled by the use of proper methods points to the view that it may also be controlled in man. This suggests the optimistic view that aggression may yet be controlled and that it is not of necessity an ineradicable part of human behaviour.

The effects of early experience

If it is argued that patterns of aggressive behaviour may be learned by the transfer of response from one stimulus to another, then it is also reasonable to assume that the opportunities afforded for learning by virtue of early experience occurring at one time of an animal's life can also affect its aggressive behaviour at a later stage.

The conditions of rearing are important in this respect and, if infants have been exposed to particular social conditions or particular circumstances during rearing, then they may well as a result show very different levels of aggression in adulthood. Young rats, for example, having been defeated early in life in encounters with trained fighter mice, no longer stand to face an opponent in subsequent aggressive encounters, but squeal and run away (Kahn, 1951). In this case early experience of defeat ensures that the animals were defeated again when they were next faced with a similar situation.

When animals are reared in isolation, not only are they denied the opportunity for social communication, but they also never experience defeat by another animal. This could have the consequence that they never learn submission, they are unsocialized, and aggression occurs at a high level. This is certainly the case with mice, and animals reared in isolation were more defensive, less willing to investigate other animals, and markedly more aggressive than others (Kahn, 1954; Hudgens, Denenberg and Zarrow, 1968).

Yet another effect of early experience upon aggressive behaviour is that of handling. Mice which have been handled during early infancy, apart from the well-known changes in emotional responsiveness, are found to fight much more rapidly when placed in an arena in pairs (Levine, 1959). Whether this result means that the handled animals were less fearful or more aggressive is difficult to say, but this result does again point to the fact that levels of aggression are considerably affected by the nature of early experience. This once more suggests that the period of infancy is one of particular sensitivity to environmental effects, and that the mechanisms of aggression also come under these influences, such that the forms and dynamics of the behaviour are radically altered.

Problems in the study of aggression

Lorenz (1963) argues that aggression comes about because of some form of internal urge to attack. An individual does not wait to be provoked, but rather the urge to fight builds up in the absence of provocation, until finally the animal seeks the opportunity for fighting and indulges in spontaneous aggression. Lorenz appears to ignore ontogenetic factors in the development of aggression – for example, the effect of learning that it is possible to stave off physical injury by attacking or at least threatening first, and the background of learning by which aggression that brought rewards in the past would increase the frequency of that aggression in the future.

Lorenz puts forward the view that the response mechanisms pre-exist and that the energy and volition bound up within them must be held in check by inhibiting forces; principal amongst these are pacifying stimuli displayed by potential victims to ward off attack.

This view, however, seems to be unnecessarily complicated. Even if it is argued, on the basis of the evidence from studies of hypothalamic stimulation and decortication, that the mechanisms of aggression are pre-formed and established and await stimulation to call them out, it is not necessary to suppose that they have an intrinsic energy of their own which is constantly leading them to seek expression. A far less complicated view would be that the mechanisms of aggression do exist in a preordained fashion but that they are switched on in response to the needs of the organism; there is no build-up of an internal urge but rather a linking of the stimulus with the pattern of aggressive behaviour. A simple analogy would be that of an engine which would be switched on or off as its function was needed. When not in use, it is not seeking actively to express itself but lies dormant and inactive until such time as its potential may be employed. If this view is adopted, then the concept of restraining forces becomes unnecessary and aggression can be seen as a function which is called out by stimuli, but not necessarily held in check by other stimuli. The stimuli used to call out the response may be those, for example, directly related to pain, or the more complex stimuli associated with frustration, with the need for self-preservation and the other primary demands of the organism.

Much discussion has centred on aggression in relation to pain stimulation, frustration, sexual rivalry and competition for food and available space. In all of these cases it is possible to point to aggression, whatever the nature of the stimulus conditions leading up to it, as potentially securing rewards which would otherwise be unobtainable or more difficult to get. In the case of pain and frustration aggression may well bring about the absence of these conditions, and this could be seen as rewarding in its own right. Aggression in this sense can be viewed as an instrumental act. This is effective in bringing about some change in the social circumstances. Even temporary changes in the level of aggression can lead an animal to win out in competition with others for food, and instrumental aggression can operate in all conditions to satisfy drives to attain goals, or to regulate the animal's behaviour in other ways. Aggression can be used at the service of any prevalent drive system. Any kind of hostile aggressive act that served to bring about satisfaction in the form of rewards in the past, could be used by the animal to serve some end on future occasions, and this would account for an increase in its frequency.

The main purpose of aggression, insofar as it is possible to talk about it, appears to be to provide a series of instrumental techniques which are used primarily for the protection of the individual and for securing its needs. It is a form of behaviour which is self-maintaining in the sense that feeding, drinking, grooming and other processes are. It is used by the animal for the purposes which are relevant at a particular time, whether this be the warding off of conditions which bring about pain and injury, the defence of territory and young, competition for sexual partners or maintaining the established food supply.

Summary

The term aggression relates to behaviour patterns of several kinds. These are attack and hostility, as well as the threat display. Threat displays stand in place of attack and can be used as a signal that future attack is probable. Definite areas of the brain relate to the production of aggressive behaviour. The release of aggression by hypothalamic stimulation suggests that the mechanisms of the behaviour are pre-formed and that they await the appropriate stimulus to call them out. The fact that aggression in these studies may not be directed to a specific object suggests that the object of aggression should concern us less than the patterns of behaviour themselves. Sex hormones relate to some aspects of aggressive behaviour. The effects of castration in the male is to bring about a reduction in the status level as determined by aggressive encounter, and the administration of testosterone in most cases acts to restore the aggressive response to its former level. There is some evidence to suggest that the hormone influence is most marked at early periods in the life of the animal.

Aggression occurs as a response to painful stimuli. There are numerous examples where pain consistently calls out aggressive behaviour. It is possible that much aggressive behaviour occurs as a conditioned response to any stimuli which have been associated in the past with pain. Painful stimuli experienced during infancy may be an important factor in the early appearance of aggression. Stimuli which originally had a neutral effect are capable of calling out aggression after conditioning has taken place.

Aggression can be easily established by rewarding animals. Pigeons can be trained to peck at other pigeons for food reward. Another important condition in giving rise to aggression in studies of learning is that of frustration. Frustrated animals turn and attack other animals. Frustration cannot be overlooked as a contributing factor in the situation where animals find themselves in competition with one another, because the possible ascendancy of one animal inevitably frustrates others.

The fact that aggression may be abolished or at least suppressed by appropriate forms of training again suggests that learning is a powerful force in relation to aggression. Whilst it is acknowledged that the mechanisms of aggression may be pre-formed and established, the stimuli which call them out need not be, and aggression can be viewed as an instrumental act which is effective in social relationships. Aggression provides a series of techniques by which the animal protects itself and secures its own needs, as well a those of the family and social group.

At the present time many investigators from diverse scientific disciplines are devoting their attention to the study of social behaviour in animals. There are, however, many changes of emphasis in this current perspective. One of these changes has been to regard behaviour as much more finely balanced, as more adaptive and malleable, and as less fixed and rigid than had previously been supposed. Many of the earlier naturalists and zoologists concentrated their attention on behaviour forms or units which appeared relatively fixed and stereotyped (Craig, 1914; Heinroth, 1910). The investigators of animal behaviour are now pursuing the more complex processes by which units of behaviour of this kind become linked into chains. An awareness has grown that, whilst the behaviour unit may be inflexible, the way in which it interrelates with other units is not so, and attention begins to be directed to these more malleable aspects of behaviour and the processes by which each individual adapts to the nature of the environment. With this change of emphasis has come the realization that, as social life is complicated and 'intricately balanced', something more than a list of straight-forward behaviour patterns is necessary if a discipline is to develop which is able to provide a full account of the nature of the communication and interaction between organisms. A comprehensive discipline which sets out to provide a full explanation of the social behaviour shown by animals must allow due weight to the subtle interactions, the continuous adjustment and the important dialogue and intercommunication which must take place, all of which relate to the learning of social responses and the learning of social interaction.

Much of the evidence points to the view that a great deal of social behaviour is learned, as well as suggesting that learning operates in the implementation of fixed-action patterns of behaviour. Learning allows the animal to be adaptable and to put the behaviour patterns in its repertoire to good use. Learning allows the animal to call out the appropriate behaviour in response to the requirements of the situation.

The view of learning employed in this chapter is one of adaptation to the environment. Learning is the process by which complex interactions with the environment are made possible. Learning is the means by which important social interactions are maintained.

The effects of early social experience

The first important point to note about the nature of social learning is that influences brought to bear during infancy can radically alter the form of behaviour shown during the rest of the organism's life. Kuo (1938, 1967), for example, carried out a series of experiments on the killing of rats by cats. Kittens which had been reared with rats failed to show the usual response of killing them and in fact lived with them quite peaceably for a considerable time. When the kittens were shown by other cats how to kill rats and to eat them, some of the kittens attempted to kill the rat in their cage, but only a few succeeded. The rat in most cases drove the kittens off by biting and by the ferocity of its attack. Kittens also failed to kill sparrows placed in the cages with them, if the sparrows were placed there whilst the kittens were still young. Later some of the kittens, but only a small number, played with the sparrows or caught them in flight. It appears from these experiments that animals can work out an amicable arrangement with animals of another species which they would normally kill, if they are allowed close contact with them during the earliest periods of their lives. Tsai (1963) also carried out experiments on the effects of early experience on social learning by housing young kittens or alley cats with white laboratory rats. These animals, which might appear to be natural enemies, could live peacefully together when young. These experiments point out the importance of early social learning and the part it plays in the regulation of subsequent inter-species differences.

The importance of maternal behaviour in the early interaction with the young has been discussed in a previous chapter. Maternal influences do bring to bear powerful forces for the modification of behaviour during the early period of an animal's life. Denenberg, Ottinger and Stephens (1962) reared animals by a process of multiple mothering, i.e. two mothers were shared between two litters, the original mother looking after the litter at one time, a foster-mother looking after it at another. They found that animals reared by this method were more emotional in adult life than those animals reared by a single mother. Maternal stimulation during infancy appeared, therefore, to be an important variable in changing the nature of the behaviour of the offspring. Ader *et al.* (1960) also showed that the effects of the interruption of maternal care may be related to ulceration of the offspring. Animals experiencing interruption of the contact with their mother during the first ten days of life were found to be much less susceptible to ulceration, when later placed in a conflict situation, than were those animals whose maternal contact had been interrupted during the second ten days of life.

The effects of social isolation during infancy have been discussed previously. These are profound, and they also point to the importance of early social learning. Rosen (1964), for example, reports that male rats reared together as a group occupy a higher position in the dominance hierarchy than rats reared individually. Reynolds (1963) found that rats reared in isolation showed very poor performance in learning to escape from areas of the cage giving them electric shock. Morrison and Hill (1967) showed that rats approached more frequently in an approach-avoidance conflict situation when they were in a group situation than when they were isolated. Animals reared in groups also showed this social effect to a greater degree. The suggestion was that the presence of other animals acts to reduce fear and that this group effect is acquired from an early age.

Smith (1969) has found that rats spend much of their time in proximity to one another and also that rats which have been reared in a colony spend a great deal of their time on a platform next to another cage in which a rat has been retained. Animals isolated from companions early in life do not show this behaviour to anything like the same extent. Smith also found that socially reared animals press levers to view other animals more frequently than animals which have been reared in isolation.

One of the most important influences within this area has been studied by Angermeier, Philhour and Higgins (1965). They found that if rats had been reared alone, the presence of other rats in a conditioned escape situation did little to reduce their fear. If, however, the rats had been reared from birth with other animals, then the presence of other animals during the conditioned escape training acted to reduce fear considerably. The hypothesis was expressed that reduction of fear seemed to have developed in social situations early on in the life of the organism. This result has the important corollary that, where socially reared animals are brought together for the performance of a particular task, they will be less fearful than if they are required to perform the task individually. This raises the important consideration in most learning situations where individual animals have been studied that the growth and development of performance may not of necessity be the result of learning the problem, but the result of learning not to be afraid when isolated and removed from other companions. This result also raises questions about the performance of animals in imitation learning experiments. Typically one animal is trained alone in the situation, and then a second animal is introduced with the first to the situation. The animal allowed to imitate the first seems usually to show much better performance and quicker learning than the animal required to learn the problem by itself. Much of this improvement in performance can, however, be ascribed to the fact that the second animal is not isolated from a companion and may consequently be less fearful than the first which has been trained in isolation. This result creates problems, not only in a consideration of the nature of imitation learning, but also in considering all the work so far performed on animals in learning situations in which the animal learns in isolation. In addition this work does point to the significance of the period of early development as a time of peculiarly intense activity with regard to social learning, a period during which influences are capable of modifying the behaviour of the organism throughout the rest of its life.

Gregariousness

Without doubt, members of many animal species show a considerable attachment to one another. Dimond (1970) investigated the behaviour which domestic chicks show when they are separated from their companions by a perspex screen. Latané and Glass (1968) also revealed in their experiment something of the power of the social bond. Rats placed together in an enclosure showed a considerable attraction for one another and spent their time in close proximity. Rats did not, however, spend their time close to other objects placed in the cage with them.

Sociability and the degree of gregariousness depends to a very large extent upon the conditions and the kind of society in which the animal has been previously living. If changes are introduced prior to testing in the nature of the group in which an animal is maintained, then this is reflected in the animal's gregariousness (Shelley and Hoyenga, 1967). It was found in these experiments on the effects of social isolation in infant rats that variability in sociability behaviour can be produced by as little as two to five days of modified social living conditions and that these effects are reversible.

Salazar (1968) also reports on the gregariousness of young rats. Isolation prior to testing considerably affects their sociability, and here again the results indicate that sociability may be considerably affected by manipulation of the social environment prior to testing. These results are interesting because they illustrate the fact that social influences change the behaviour of animals over the short as well as over the long term. These results indicate the importance of gregariousness as a powerful tendency in social-living animals. The possible reasons for gregariousness are discussed later.

Social facilitation

There are many effects upon the behaviour of animals
which result from the presence of companions during the
performance of a particular task. The presence of another
individual may produce an increase in individual activity
and possibly an improvement in performance, which can
be described as social facilitation. Welty (1934) trained
goldfish to swim through a door in a dividing compartment
in their tank to obtain food. The goldfish was required to
swim backwards and forwards from one compartment to
the other in order to get the food reward. The principle
measure of learning was the speed with which the animal
moved from one compartment to the other to obtain food.
Fish in groups learned this task much more quickly than
individuals, and if one fish had been trained previously
as a leader, then the untrained fish tended to follow this
leader with the result that they also learned very quickly
how to get food.

One of the most striking of the early observations of
social facilitation concerns feeding. Bayer (1929) found
that hens which had been allowed to satiate themselves
with food, began eating all over again when other hungry
animals were introduced into the situation. If three fresh
hens were introduced, then previously satiated animals
ate more than if one hungry hen had been introduced. Tolman (1965) showed that actual physical contact between
one chick and another was not necessary in order for facilitation of feeding to take place, providing the animals could
see one another through a screen. The effects of the companion appears under these circumstances to reduce the
emotional behaviour shown by the chick in entering a
novel situation. The companion is part of its past experi-

ence and so provides a familiar element in an otherwise strange and presumably frightening new test environment.

Social facilitation of feeding has also been found in rhesus monkeys (Harlow and Yudin, 1933). A considerable degree of extra feeding was elicited in social situations. The results suggested that monkeys are peculiarly susceptible to social facilitation of this kind, and that it occurs even when there is no risk of competition, in situations, for example, where the monkeys are fed in separate cages one foot apart. James (1960) studied the development of social facilitation in puppies. These animals, with one exception, did not show facilitation when fed with others on the first trial, but developed it subsequently. Once the effects of facilitation had appeared, the puppies always consumed more in the social situation.

A number of studies have been undertaken of social facilitation in laboratory rats. These have not always proved successful and in some cases for very good reasons. Levy and Bevan (1958) failed to demonstrate any effect of social facilitation with regard to audiogenic seizure in rats. Audiogenic seizures are a form of behaviour which is elicited by intense auditory stimulation. The animal appears to have a seizure and shows typical patterns of muscle spasm. This is not a form of behaviour which could be considered to be social in any accepted sense, and it is not surprising that the presence or absence of cage mates had little effect upon it. Also Holder (1958) carried out a series of studies in which the presence of or absence of one rat was associated with reward or lack of reward for another rat. The nature of the reward and its association with a stimulus rat was important in determining whether that stimulus rat produced social facilitation of learned response. The rats were tested in a situation in which they were required to press a lever to release the stimulus rat from a box before either of the rats could obtain any food reward. If the stimulus rat had been previously associated with the absence of reward, the rat under investigation was slow to press the lever to release it. If, however, the stimulus rat was one that had previously been associated with reward, then the rat operating the lever pressed readily to obtain its release in order that both animals might obtain food reward. Social facilitation in these experiments appeared to be a question of the degree to which stimulus rats have been associated previously with reward.

Social facilitation is a widely reported phenomenon. Simmel (1962) found that social facilitation takes place with regard to exploratory behaviour. A rat was allowed to explore an object, and then a fresh rat or one which had previously explored the object was introduced to it. The rat showed a good deal more additional exploration of the object when placed with a rat that had never explored it before. Pishkin and Shurley (1966) found that social facilitation of operant bar-pressing rates was increased significantly if the animals had suffered antecedent sensory deprivation. It is not clear exactly what the reason for this finding is, but it seems likely that periods of sensory deprivation enhance the stimulus value of other animals and thus lead them to have a greater effect as stimulus objects.

Klopfer (1961) has pointed out that a distinction may be drawn between social facilitation and true observational learning. The presence of other animals facilitates performance at the task whilst they are there. Klopfer supposed the effects to be transitory, which means that facilitations is not permanent and that the effects are no longer present when the companion is withdrawn. Observational learning on the other hand does persist over time, and the learned response is still present after the removal of the animal which, however, unwittingly acted as the teacher.

Not all the effects of social interaction, however, need be expected to take the form of facilitation. On some occasions the presence of other animals interferes with the task in hand; on other occasions the form of social stimulus provided by the other animal is irrelevant and has neither facilitating nor interfering effects. The results obtained depend very much on the nature of the task and upon the type of response demanded from the animal. Many enquiries do report social facilitation, however, and providing the task is one which can be performed without distracting interference, does not itself produce high levels of anxiety and fear, and providing the task is one on which social factors can operate, then it is not difficult to observe these effects.

There are several reasons which could be put forward as an explanation of social facilitation.

First, the performance of an animal may be more disrupted if it is placed by itself than if it is placed with companions. The isolated animal may show emotional disturbances characteristic of fear. It may show freezing and immobility or other responses which interfere with its performance. Latané and Glass (1968) found that rats were far less fearful when they were tested together than when they were tested alone.

Secondly, social facilitation could result from the excitation which one animal induces in another. Each animal acts as a stimulus to other animals. Group-reared animals provide a powerful stimulus to each other. Strange animals of the same species provide a stimulus through species identification. Each animal is a source of movement and activity which differentiates it from the rest of the environment. Each animal could form a stimulus and a source of excitation to other animals. Animals performing a task in the presence of companions are more activated than those performing it in isolation.

Wheeler and Davis (1967), for example, found that if rats were required to delay response for periods of up to ten seconds before getting food reward, then isolated rats could be trained on this task to a level of stable performance. If, however, animals were required to perform in the presence of other animals they no longer reacted so successfully to the constraints placed upon them. They began pressing the lever to obtain the food before the ten-second period had elapsed. This finding suggests that the presence of other animals increases arousal, and in this case the stimulating effects of other animals, because of the nature of the task, interfered with proper performance.

A third factor in social facilitation concerns leadership. If one animal acts as a leader, an animal following this leader may be induced to perform a task in a satisfactory way because its attention has been directed towards the essential characteristics of the task. If one animal starts to feed, it directs the attention of the others towards the nature of the food. The goldfish placed in a shuttlebox with a trained leader learns to respond quickly because of its association with the leader. The stimulating properties of the leader induce following behaviour in the other animal. Social facilitation in this case comes about through imitation as well as through the tendency of these animals to remain together; these questions will be discussed in a later section of this chapter.

Patterns of social reward

It was seen in previous chapters that one animal is capable of acting as a reward to another animal. A sexually deprived rat will, for example, press and continue to press a lever to obtain receptive female rats until it is sexually exhausted. A domestic chick will work to open a window in order to view an object to which it has previously become imprinted (Peterson, 1960). A female monkey will spend time working to open a window in order to view an infant monkey. Butterfield (1969) has investigated pair bonding in zebra finches. One finch on lighting upon a particular perch causes a light to be turned on in an adjacent cage which illuminates the animal placed in that cage. One animal will work more effectively to view its own mate in preference to other animals. The fact that animals can be induced to work effectively in order to view other animals, particularly those which bear some relationship to them, indicates that contact of this type is reinforcing and that one animal is capable of acting as a reward to another animal. Contact, vision, or the sound or smell of other animals can each act as a reward for behaviour.

The reinforcing properties of one animal for another have been studied by a number of authors. Angermeier (1960) allowed rats to press a bar to obtain ten seconds either of visual or physical contact with another rat. If the rats had been reared together they quickly learned to press the bar; if, however, the rats had been reared separately, they were not at first proficient at this task but rapidly became so as they grew more and more accustomed to the contact. In a later experiment Angermeir (1962) trained adult rats to press a lever. After they had done so they received visual reinforcement from another rat, a domestic chick, or a chick plus a 100-watt light. Adult rats pressed consistently to receive social reinforcement from the other rat, but novel or noxious stimuli resulted in a considerable depression of bar-pressing performance. Thach (1965) also found that one animal would work consistently in order to view another animal and that this response was soon extinguished when the stimulus animal was withdrawn, although occasional bar presses persisted even after the extinction phase. This could suggest long-term as well as short-term extinction effects. The animal may press the bar over the long term in the off-chance that the previous conditions will have been reinstated.

Gallup (1966) also pointed to the significance of social stimulation as a reinforcer of behaviour. He found that pigtail monkeys will readily work, not only to view one another, but also to look at themselves in a mirror. Gallup required the monkeys to open a door in order to look at themselves. This allowed them a period of 60 seconds of inspection. A Japanese macaque monkey raised with pigtail monkeys did not respond in a positive fashion to the mirror. Instead, it tried to close the door once it had been opened. The suggestion arises in this case that, as the Japanese macaque had been raised with pigtail monkeys, its own image would be totally unfamiliar to it, and would in fact form something beyond the limits of its experience. In attempting to shut out the image it could be responding with fear equivalent to that in response to an animal of a completely strange species.

Thompson (1963) demonstrated the importance of visual reinforcement in modifying the behaviour of the Siamese fighting fish. He found that these animals will learn to swim through a ring, if immediately afterwards they are allowed for a brief period to see themselves in a mirror. Goldstein (1967) also found this effect when studying the same animal. He demonstrated, too, that if the mirror image is presented independently of the type of behaviour which the animal is showing, then whilst there is a slight increase in the operant response this is not a sufficient increase to provide an explanation of Thompson's original finding.

Much of the social behaviour must relate to the fact that one animal finds the presence of another animal rewarding. Within this reward system must surely lie the roots of gregariousness: aggregation, flock formation, and social grouping. It is a matter of speculation at present as to exactly why one animal should form a reward for another, but surely one of the important factors must be that of the animal's early experience. The young reared in a family group interact with one another from an early age. The family group provides food, comfort, shelter and warmth, all of which the animal presumably learns to associate with members of its own species. There are opportunities for learning over and above this, however. Each animal learns to communicate with others. The suggestion arises that a sophisticated behavioural language is learnt from the earliest period of the neonate's existence, and this means that in addition to other forms of association each animal acts to the other as an important communication source. Contact with other animals is rewarding from this point of view because it allows communication to take place.

Finally, the presence of companions may act to reduce fear, because companions form a familiar stimulus in what may be otherwise an unfamiliar environment. In seeking the companionship of others an animal could move from the unfamiliar to what is known and what accords with previous experience. Movement towards the familiar may explain something of the attraction which most animals of the same species maintain for each other.

Cooperation and competition

One of the striking features of many social organisations amongst animals is the degree to which in these societies one animal is prepared to cooperate with another. Cooperation is noticeably demonstrated in family life, and examples of it are found in such early interactions as occur in play, sexual behaviour and group formation. Cooperation to achieve some particular end is characteristic of many animal societies. It has been pointed out that often the combative and destructive aspects of evolution have been over-emphasized at the expense of factors such as cooperation, protection and social facilitation, which all tend to limit combat and destruction between animals. Carter (1951) has also pointed out that whilst competition may be a necessary feature of natural selection, not all interactions are competitive and active cooperation is 'a real fact of natural history'.

Several investigators reported early on that rats appeared to cooperate with one another under certain circumstances. Mowrer (1940) showed that one rat will work in order that both it and other rats may feed. Daniel (1942, 1947) also showed that rats would learn to take turns at sitting on a platform in order that all the animals could feed without being interrupted by electric shock.

Skinner (1962) found that it was possible to obtain cooperation from animals in the operant situation. In Skinner's experiment there were two pigeons which were each required to peck at a series of buttons. A pair of buttons of this series was effective in producing food reward. Both of these buttons had to be pressed, one by one pigeon, the other by the other. A high degree of coordination of response was achieved by the animals in this situation and, after training, each bird became highly sensitive to the behaviour of the other and showed a strong tendency to imitate the other with regard to other aspects of behaviour, such as drinking.

Church (1959) found that rats were to some extent responsive to discomfort in other animals. A rat pressing a lever for food would stop pressing, or at least reduce the rate of pressing, whilst another rat suffered pain through electric shock. If shock administered to one rat was used as a conditioned signal that shock was to be administered to another, then lever pressing by the second animal showed a marked decline in response to the shock administered to the first animal. This conditioned response in one to the shock administered to the other persisted for several days, but the animals quickly adapted if shock was administered to another but they themselves received no shock.

A controversy has developed as to whether experiments of this kind represent altruism or some other factor in rats. It has been argued by Lavery and Foley (1963) that the results of these experiments do not represent altruism at all but simply the effects of activation. They found that, if rats are exposed alternatively to white noise or to the recorded squeals of another rat, then they will press a lever not only to turn off the squeals of a rat, but also to turn off the white noise. They argued that in both cases the stimulus induced the rat to show increased activity, which resulted in it pressing the lever which turned off the stimulus. It is possible that both white noise and the distress calls of another rat are equally unpleasant to a rat and so it turns both stimuli off. This is an anthropomorphic remark, however, and we are in danger of interpreting the situation in human terms; but the argument still holds that, because the animal is prepared to turn off white noise, this does not of itself mean that the animal turns off the squeal of another rat for the same reason, or that it is not capable of showing altruism in doing so.

Rice and Grainer (1962) took up this problem in a further investigation. In this study an albino rat was found to press a lever to lower a distressed rat hoisted above the cage. A plastic block was not lowered to anything like the same extent although it was of roughly the same shape and weight as the rat. Rice (1965) also found that one rat would release another rat from water immersion and again described this as altruistic behaviour. Guinea-pigs on the other hand did not show altruism in this sense, which is surprising in animals which appear generally to be highly sociable. Horel, Treichler and Meyer (1963) managed to obtain some degree of cooperation in their animals by virtue of the fact that they used primates, and one monkey actively coerced another monkey to perform a response which would bring both monkeys food. Two monkeys were first trained independently to turn a handle in order to obtain food. They were placed, both together, in the cage. The cage contained only one handle, and the animal turning this most frequently was assumed to be the dominant animal. The apparatus was then fixed so that it no longer worked for this animal. The only way this animal could get food was to snatch it from its partner after the latter had turned the handle. The partner was now fed, and as a consequence was no longer interested in turning the handle to obtain food. The dominant animal became hungrier and hungrier. A few of the animals under these circumstances were observed to coerce their partners. They oriented them and pushed them forwards towards the food-well. One of the pairs of monkeys pulled the hind leg of its partner and thus caused it to move the handle releasing the food because it had been tipped off balance. This kind of coercive behaviour is, however, relatively rare.

Hanson and Mason (1962) investigated one aspect of co-operative behaviour in monkeys by training them to depress one of two levers to produce food. Each monkey was found to show a particular preference for the use of one lever rather than another. In the second part of the experiment, every time the preferred lever was pressed another animal in the same cage was given an electric shock. This had quite a marked effect on the rate of lever pressing. When each lever press acted not only to provide food reward for one animal but also to give electric shock to another, the rate of lever pressing declined. Attention was now directed away from the lever producing the shock towards the other less preferred lever. This lever did not pair food reward with shock to the other monkey, and it was now frequently used by the operator. Wechkin, Masserman and Terris (1964) confirmed the finding that a hungry rhesus monkey will avoid securing food by lever pressing if at the same time this subjects another monkey to electric shock. The hungry monkey sacrifices the satisfaction of its own hunger in order to avoid subjecting another animal to pain.

The question of altruism in animals is a very vexed one. If it were possible to explain this type of behaviour with reference to simplified concepts, for example that of activation, or reward to the animal behaving in a socially helpful way, then it would be necessary to do so. The attempts to explain altruism as nothing more than the effect of extra stimulation by one animal upon another, have been necessary but, at least as far as primates are concerned they do not really provide a satisfactory explanation of all the evidence which is available. An alternative explanation is necessary, and it could be supposed that there is a premium upon cooperative behaviour from early infancy. If the infants cooperate with the mother within the litter, then they are fed, cared for, protected and kept warm. If they are uncooperative, they may well be punished.

In play animals learn the patterns of cooperation. They learn how to behave socially without exceeding certain limits of aggression, and cooperation of this kind brings its own rewards. The non-cooperative animal evokes aggression in its turn and rejection from the other group of animals. Perhaps the most obvious example of cooperative interaction in adult social behaviour is that of sexual response, in which a complicated chain of interaction takes place. Another example is that of parental behaviour where care is solicited and given in the interaction between the parents and the young. There are many instances of this kind, and the suggestion which is put forward here is that this behaviour relates to learning in early infancy and to the behavioural dialogue between parents and their young, as well as between one infant and another.

There are obvious differences between species in the degree to which they cooperate on various kinds of learning tasks. Species low on the phyletic scale often fail in cooperative tasks they have been set. They show, however, a considerable degree of cooperative behaviour. This fact suggests that the tasks they have been set do not always match the form of cooperation which they are best capable of showing. In many cases the tasks demanded of them have been 'intellectual', with the result that those animals having most difficulty in learning a new task would have most difficulty in learning a task in which the elements were social rather than, for example, the turning points on a maze or the coloured compartments in a shuttle box.

Competition has been regarded as the foundation upon which the selective mechanisms operate during evolution. Our concern here is not with studies of dominance and leadership and aggression *per se* but with competition in experimental learning situations. Skinner (1962) began a study of competition in experimental conditions when he required two birds to play a game in which each was opposed to the other. This game resembled 'ping-pong'. The task presented to the pigeon was to make sure that the ball was returned to the opponent's court. The pigeon was required to peck the ball, making it rise over a shallow incline and then down to the opponent's side. If both birds were rewarded, they maintained a coherent game. If on the other hand one bird was rewarded at the expense of the other, then the pecking response would be extinguished in the latter and the game would cease. If, however, the body weights were changed and the more unsuccessful bird became lighter and the more successful bird heavier, the game could be reinstated.

Baron and Littman (1961) describe a competitive situation wherein two hungry rats were placed in an operant situation with only one lever and one food trough. The lever and the trough were placed at a distance from one another in the test cage. Rats seem to respond in one of two ways. They either press the lever and move quickly to the food trough to grab what food is available, acting as a worker, or they remain passively by the food trough waiting in a dependent state for the other animal to press the lever and provide them with food. These latter animals develop a parasitic mode of life in relation to the other animals, which probably relates to the dominance shown in the lever-pressing situation. Oldfield-Box (1967) found that a worker–parasite relationship developed when two animals had been separately trained in an operant situation, and they were now placed in an apparatus with one lever and one food tray. One rat did most of the work in pressing the lever, the other rat consumed the products. Group-trained rats, on the other hand, gave rise to a much more flexible arrangement: one rat pressing some of the time, another rat the rest of the time.

Competition between one animal and another can be responsible for a change in the rate of lever pressing. Rats may be trained in adjacent boxes from which they can obtain information about each other. They soon learn to press the lever to obtain food reinforcement. If now they are made to compete with one another, it is possible to bring about a change in the rate of lever pressing. After every 15 seconds, in one experiment, only the animal scoring highest received food reinforcement, and this acted to increase considerably the rate of lever pressing in both animals. In another experiment (Church, 1961), only the animal responding least within each 15-second period received food reward. In both experiments the animals responding quickest initially were the ones which showed the greatest subsequent modification in behaviour. The results should be interpreted with caution, however, because it is possible that social interaction could be of only limited importance within this context. Animals were being reinforced for high rates of response in the first experiment and low rates of response in the second. It is possible that the same results could have been obtained on the basis of response criteria alone, and that social factors are not as important as had been supposed.

When animals are paired in a single cage they show little aggression towards each other under normal conditions of training. When, however, the responses of one animal are extinguished whilst the responses of the other are rewarded, aggression is observed. The animal whose response has been extinguished tends to interrupt the performance of the other. When the responses of both animals have been extinguished the greatest degree of aggression of all is observed (Davis and Danenfeld, 1967). These results suggest that frustration in the extinction situation relates to aggression. They also point to the importance of competition for the existing food resources in determining the type of behaviour which one animal shows to the other.

Kanak and Davenport (1967) literally required rats to race with one another in order to get food reward. Competition between animals led them to increase their running speed much more rapidly than animals not in competition. Those animals which begin to lose out in competition quite rapidly show extinction of the running response. Kanak and Davenport regard competition as being in the category of an acquired drive which combines with certain aspects of primary drive. This view of competition is very different from the one which supposes that there are built-in mechanisms of behaviour giving rise to the ascendancy of one opponent over the other. The view expressed is one whereby competition is a learned pattern of behaviour which comes about because it is rewarded. When competition ceases to be rewarded as such, then it extinguishes and no longer appears in the behaviour repertoire. Competition, therefore, appears as a learned mode of behaviour, which is over and above that of the life contest of organisms deprived of some aspect of their primary drives. This seems to be begging the question, however. If we make animals compete, they are not just racing one another; they are doing so because they are hungry and need food to survive. Primary drives are involved and the competition is for the satisfaction of primary drives. If the response extinguishes, then that animal has lost the competition and, if confined solely to the runway, would starve under these conditions. In a sense the competition is artificial in that both rats are placed hungry into the runway and only one gets the reward. Under natural circumstances a defeated rat would look elsewhere or, if sufficient food were available, would obtain food after the other had fed. but this experiment does illustrate the nature of the behaviour in this situation. It illustrates the dominance of one animal and the rather hopeless withdrawn condition of the loser, who would, if the experiment persisted, be in serious danger of his life.

The experimental investigation of both cooperation and social conflict reveals that many animal species are capable of both. The question becomes, not whether an animal is capable of cooperation or of competition, but what the circumstances are which give rise to cooperation and to competition? Both appear, at least to some extent, to be adaptations to the nature of the task which the environment is setting. Both can be seen as a response to the demands of the environment. Both can be seen as a function of the reward systems prevalent at the time and as a function of the early experience which the animal has received within the infant and family group.

Observational learning

There are many instances in animal life whereby learning comes about because one animal observes another animal performing a particular pattern of behaviour. Its own performance is then enhanced. This has been described as observational learning, and this term covers all cases where observation of the performance of one animal facilitates the learning of a task by another animal. In one sense all learning is observational whether it occurs in a social context or not, but the term is used here to apply specifically to social learning. It embraces the area of imitation learning but does not rule out consequences of social interaction, such as social facilitation. Some of the older writers talked about the 'instinctive imagination'. Washburn

(1908) uses this concept to explain the behaviour of a group of chicks when they learn to drink from a tin of water. When one chick has discovered the water by casual experience, pecked at it and drunk, the rest run up and also start pecking and drinking. Individuals in this case learn by observation where to peck and what to peck at. This kind of behaviour does not involve a new motor organisation; it does not mean the learning of a new response but simply a social focusing towards the source of the stimulus.

Another more complex example of observational learning occurs in the report by Angermeier, Schaul and James (1959) that avoidance conditioning can be established by social training in the complete absence of electric shock. One group of rats had been trained to run into a compartment when a buzzer sounded, indicating that if they remained where they were they would get an electric shock. If fresh rats were introduced into this situation with the trained rats, then although shock was not given to them they very rapidly learned the task and ran into the safe compartment with the trained rats. Later, when they were placed in the avoidance box by themselves, they showed a fully conditioned response to the buzzer, which had come about by virtue of the social training which they had received. Mowrer (1960) regards observational learning of this kind as coming about through a process of empathy, whereby one animal responds to the emotional reactions of other animals. The animal empathizing must, he argues, not only be keenly aware of the responses of the other animal, but also must (according to Thompson, 1958) be able to act as if it were in the place of the second animal. We may, however, doubt that this last condition is strictly necessary, but certainly it would have to be inferred that one animal was highly responsive to the behaviour of others and that the emotional state of one animal can be communicated to others.

Gilbert and Beaton (1967) established quite complex chains of behaviour in male black-hooded rats. Each rat was required to respond to a light above a speaker by switching it off; this produced a 2,000 KHz tone. The animal then had to pull a trapeze which turned the tone off and illuminated a light above the food-cup lever. The animal subsequently had to press the lever to obtain food. When an animal was placed in an adjacent transparent cage and required to perform the same ritual, one rat in learning this task began to adopt the mirror-image posture of the other rat as it went around its cage getting food reinforcement. This mirror-image following led to a very rapid learning of the task and the achievement of the food reward. Carson (1967), again using laboratory rats, found that they learned to press a lever quite rapidly to obtain food reward if they were able to observe other animals performing the same task. It is suggested that this may be a useful technique for training animals to press the lever in the first place, so that other investigations can be made upon their behaviour. Cats were studied in a similar situation and were found to be very proficient at observational learning. Animals higher in the phyletic scale appear to be adept at acquiring responses in the presence of companions. The primates appear to be particularly good at observational learning. Darby and Riopelle (1959) found quite an enhancement of performance when one monkey was allowed to watch another monkey only for a single trial. Studies have also been undertaken of imitation in avoidance learning by monkeys and apes. This area of study would seem to be very important, because the ability to learn to avoid sources of danger by observing the behaviour of another animal would have great survival value.

Presley and Riopelle (1959) trained one monkey to avoid shock by leaping over a barrier within four seconds of the onset of a red light. Another animal watched the training from an adjacent compartment. The observer was then placed in the training box and showed very rapid learning of the avoidance problem.

Several different theories of observational learning have been put forward to explain this phenomenon in animals. It is not possible to review all the evidence relating to each theory, and so only a brief statement of the theory will be given.

The first view is the reinforcement one, which supposes that any behaviour which is reinforced tends to be stamped in. Orientation towards and observation of other animals is a basic response in the repertoire of the animal, which when it occurs brings its own rewards in terms of primary reinforcements such as food and water, and perhaps in terms of less closely defined rewards such as the reduction of fear. A pattern of behaviour which is found to bring rewards becomes established as a more or less permanent habit, and thus one animal learns to observe the responses of others.

Behavioural contagion has been put forward as an explanation of the results obtained in observational learning studies. Because one organism is performing in a particular way, other organisms experience a reduced threshold for this form of behaviour and so the behaviour spreads throughout the group. It is not possible, however, to speak of behavioural contagion in the absence of those possible circumstances which may have given rise to it. Behavioural contagion could be regarded as behaviour serving the purposes of group organization. Animals not conforming to the group pattern stand the risk of not receiving the rewards of belonging to the group, and it is not possible simply on the basis that one pattern of behaviour communicates itself between group members to suppose that there are no overall influences directing this form of behaviour, which lie beyond the simple fact of contagion or facilitation.

Mowrer (1960) proposed a general theory of imitation by which emotions mediate learning. Mowrer defines empathy as the imitation of affect. He supposes not only, therefore, that imitation takes place by the simple observation by one animal of the behaviour in another, but also through a state of regard for the feelings or an interpretation of the state of experience of the other. He quotes as evidence the work of Church (1959), which showed that rats show fearful behaviour in the presence of other fearful rats. However we interpret these results, we have to beware of interpreting animal behaviour in anthropomorphic terms, and Mowrer is in great danger of doing this. He does call attention to very important processes in the communication of emotional state from one animal to another and emphasizes social rewards as a means of promoting learning.

Summary

It is obvious from this chapter that animals in many cases exert a considerable attraction for each other. Socially living animals aggregate together in groups. They are prepared to learn various tasks to perform in specific ways and to carry out work in order that they may associate with others animals. Animals on many occasions are able to profit from the experience of others. In observational learning, experience can be gained by observing the solution to a particular problem demonstrated by another animal. In addition, many of the higher mammals seem to show a kind of altruism whereby they are sensitive to the needs and demands of other animals and are prepared to help them should the demand arise. In addition to showing specific forms of conflict and competition the ability to co-operate one with the other has been demonstrated on a number of tasks. The question now arises as to the means by which this type of behaviour comes about. The suggestion arising from this chapter is that, whilst many of the individual patterns of behaviour are genetically predetermined, in that they form the repertoire of what can be put into operation, nonetheless many aspects of the social interaction whilst using these patterns would appear to be something that is learned. It is thus suggested that social interaction is something which is strongly reinforced during infancy. The relationship with the mother, particularly in mammals, is a highly reinforcing one. The interaction with other infants in play and social life is also highly reinforcing because of the conditions which pertain to it. It is perhaps a mistake, however, to think that learning can occur only when obvious reinforcements to primary drives are present. Much of the acquisition of social response would presumably occur through learning of a very different type from this; learning through perceptual registration, exposure learning, communication learning where a dialogue is established in behaviour between one animal and another, all contribute to the ultimate development of social relationships. Having established these in early infancy it is not surprising that their effects persist into adult life, because so much behaviour depends upon the interaction of one animal with another, and so many facets of life are bound up with this interaction.

Then again, the day-to-day interaction of one organism with another in adult life is a continuing process, a developing and expanding relationship. Each animal exerts an important impact upon the other, just as the nature of an animal's environment affects the way it behaves. Each animal certainly learns about its environment; it would be surprising if it did not equally learn about the companions with which it lives in that environment.

7   Status Relationships

One of the most frequently described and well-attested facts in animal behaviour is the dominance which one animal exerts over another. Animals when they encounter one another often behave in a way related to their relative status. Two strange adult animals meeting for the first time may often find themselves in the centre of a competition. They become involved in a ritual duel, and they may show either mild or severe fighting. As a result of this initial encounter one animal may lose and the other emerge as victor. After this the behaviour of both participants is to some extent predetermined. Their subsequent encounters take on the nature of a habit. The ascendant animal threatens and the other animal quickly retreats or submits. As time goes on, smaller and smaller signs on the part of the dominant animal may be sufficient to maintain the status quo. Conflict is often resolved by this means, which can lead to peaceful forms of behaviour and tranquil co-existence. Severe fighting becomes less and less necessary except in the face of a direct challenge. Ultimately, the relationship can be preserved by minimal signs from the dominant animal, which stand in place of severe fighting.

Social Rank

Some of the first work on social dominance was carried out by Schjelderup-Ebbe (1935). This work describes the formation of the peck-order in domestic chickens. When previously unacquainted birds are placed together in a pen, fights occur, each bird engaging in encounters with its neighbours. The winner of the encounter establishes dominance over the loser. As part of the benefits of status the winner may on subsequent occasions peck the loser without being pecked in return. The loser usually submits to the dominant animal and wherever possible avoids a confrontation with it. Dominance may not be completely settled at one meeting, and subsequent encounters may be necessary before the issue is finally decided. On subsequent meetings, however, once the ascendant status has been established, threat alone by one animal may be sufficient to send the other scurrying. Once established the dominance–subordination pattern shows remarkable persistence. This has been confirmed by Allee *et al.* (1939) and Collias (1943).

There appear to be several major factors which determine social dominance. First, one animal may not only be bigger, but also physically stronger than another animal, and thus enters any encounter at the beginning with a considerable advantage. Secondly, one animal may be less fearful than another. If one animal is not frightened and the other is, then the former is at a considerable advantage. Even if both animals are frightened and one recovers more quickly from the effects of fear than the other, it is at an advantage.

Thirdly, the location affects the outcome of a contest, because generally if one animal is on its home area it stands a far greater chance of success than on strange territory. Finally, there are a variety of other conditions which affect the place of an animal in the dominance hierarchy: such things as general health, fatigue, concentration of sex hormone and sexual condition, as well as possibly resemblance to former victors or vanquished in previous encounters of the animal's experience.

Social dominance may be measured in a variety of ways. The most common way of determining it in birds is through the establishment of the peck-order. The birds have to be identified as individuals, by dyeing or labelling them in various ways. The interactions within the flock are then measured by observing those birds which peck other birds and those which submit to being pecked.

Another method is to place the birds in pens two at a time, separate from the flock, in order to observe the initial encounter and decide which of the pair is the dominant animal. It is also possible to use panels or teams of birds against which the individual is matched to determine his ultimate position in the hierarchy.

Factors in maintaining social rank

Schjedlerup-Ebbe (1935) described what happens when a strange animal is introduced into a group having already established a well-defined peck-order. The newcomer is not at first made welcome. The resident birds threaten and attack the newcomer and harass it in a variety of ways. The newcomer may take a stand and put up a fight, but frequently it surrenders without any conflict and tries to get out of the way of the other animals. After a period of time it gradually becomes integrated within the flock, usually at first at a low level, reaching the level or status due to it over a period sometimes of several months.

Seniority with regard to the flock is important to some

extent in determining the ultimate social rank. Guhl and Allee (1944) periodically removed one hen from a flock at the same time as adding another. The new hen usually began at the bottom of the hierarchy, but as its period of residence with the flock increased, so its relative status improved. The longer it remained with the flock, the more dominant it became. Fresh animals arriving low in the hierarchy acted on their arrival to elevate established hens through the hierarchy. Some individuals, however, failed to rise very high, whereas others rose very quickly to the top within a few days, so that whilst seniority with respect to the home cage had a marked effect, it could not entirely override the effect of the individual characteristics of the animal.

The time taken to maintain a particular position within a flock may be relatively brief. Douglis (1948) found that part-time members of organized flocks could easily maintain their position within the flock if the time they spent with it was only one hour every other day.

It is argued that when the superiority of one animal over another has been established, hens show a great deal of toleration in submitting to their superiors. It is this submission that allows the animals to live in social groups without the rapid dispersion which might be expected in animals less able to solve the problem of aggresive encounters.

Dominant–subordinate relationships have been extensively studied in rodents. Both rats and mice have been required to take part in paired encounters of various kinds. Lindzey, Monosevitz and Winston (1966), for example, studied dominant behaviour in two inbred strains of mice. Highly stable strain differences in dominance and aggressive behaviour appeared. These authors used a food-competition situation as well as the tube-dominance test, in which two animals are placed in a tube and one animal will have to move backwards to allow the other animal to pass. Whilst food dominance correlated highly with measures of aggression, the tube-dominance test unfortunately failed to do so. Baenninger (1966) also reported highly reliable dominance orders between individuals in groups of rats. Whilst the number of aggressive encounters decreased with age, the stability of the hierarchies, when once established, remained unchanged.

Kahn (1961) pointed out that training may play a part in determining whether an animal becomes a dominant or a submissive member in a social group. Male mice trained to be aggressive show more sexual approach to a female in oestrus than animals trained to be submissive. The latter tended to be pursued by the females. This experiment points to the effect which training for specific individual characteristics may have upon the organism's behaviour.

One factor which may be important in relation to whether an animal is dominant or submissive with respect to others is the degree to which it is prepared to show exploratory behaviour. Lester (1967), for example, found that dominant rats were far more prepared to show exploratory behaviour in a 'Y' maze although there appeared to be no differences in activity or defecation between dominant or submissive animals.

Dominance orders in dogs appear to be established from an early age. At six weeks of age playful wrestling is common between littermates, but after that time the animals can on occasion give one another painful nips, and truly aggressive encounters can occur from about the ninth week onwards (Fuller, 1953). These attacks have serious consequences for the defeated animal if they are not strictly controlled. It appears likely that dominance–subordination relationships may be established in dogs from this period, and that the particular period is an important one in the establishment of this kind of relationship. Dominance relationships may make their appearance with regard to feeding. If the puppies are fed from a single pan, a dominant animal may so unmistakably exercise his superiority that another animal becomes undernourished. This marked inequality of the food rations is a feature of the situation (James, 1949). Dominance appears to be established from a relatively early age after a series of tests of strength. The heavier dog appears to have a considerable advantage when males are involved in encounters, but this factor appears to be less important with regard to females. Dominance relationships are by no means always dependent on overt fighting, and there are many signs which indicate dominance or submission to other animals without the necessity to fight on each occasion. The dominant animal stands in an erect rigid posture with a raised tail, whereas the submissive animal lowers its tail, draws it tightly between its legs, and may crouch or roll over on its back. These patterns of behaviour have been described as characteristic of wolves as well as of domestic dogs (Lorenz, 1952).

Considerable research effort has been devoted to a study of the causes of aggression between primate social groups because, of course, in an evolutionary sense they relate so closely to man. One important aid in understanding the patterns of aggression is the investigation of the relationship of dominance and submission. As a first step in the question of whether these findings gathered from lower organisms also apply to man, it is necessary first to ask whether the findings also apply to groups of infra-human primates. The answer seems to be that to a large extent they do. Leary and Moroney (1962), for example, showed that an animal occupying a particular cage which has become established as its home cage will generally exert dominance over a strange animal introduced into the cage at a later date. This relationship, once established, seems to persist at least over the course of the first week, which was the period of observation reported in the experiment. Bernstein (1964) also reported a relatively fixed dominance relationship with respect to the principal monkey in a rhesus group. The dominant male was removed from the group for a period of one month; after the removal other males took over the topmost positions. Following his return, however, the originally dominant animal still retained his superior position in the hierarchy, although it now had greater demands placed upon it in terms of the activity necessary to retain its position. It was required to be more socially active, whereas the social activity of the potential challengers was found to decrease.

Miller and Banks (1962) devised a method of studying dominance behaviour in primates. They placed two animals together in a single cage. In this cage was a perch which was big enough for only one monkey. The floor was electrified and, in order to avoid getting electric shock, the monkey was required to climb on the perch. One animal almost invariably won out over the other in this situation. The dominance relationship thus produced seemed to be remarkably stable over time, i.e. the same monkey always occupied the perch at the expense of the other animal.

Biernoff, Leavy and Littman (1964) report reliable dominance orders in food competition, using the Wisconsin General Test Apparatus. A shutter was lowered in front of two animals in a cage. When the shutter was raised, the animals could see a peanut which had been placed outside their cage. The most dominant animal was defined as the one most consistently gaining the nut. This method was also extended to studies of lever pressing in order to gain food. Plotnik, King and Roberts (1965) also report a stable linear social hierarchy in squirrel monkeys. They observed these animals under different conditions:

> The animals were required to compete in the home cage in order to get food, they were required to compete in a shuttle box for food, they had to compete to avoid shock in a shuttle box and finally as a control the aggressive behaviour of pairs of animals was observed when they were living in the home cage in the absence of competition for food or the need to escape from electric shock.

The hierarchy relating to these procedures was highly stable and persisted unchanged for periods of up to and including four months.

Boelkins (1967) pointed out that in many of these studies of social dominance in primates the 'group' exists only in that the animals come together in the test situation, but animal social groups usually form a cohesive entity which is distinguished by its stability and by the reciprocal relationships existing between the members. Boelkins therefore used previously established groups of animals in his investigations. Two groups of macaque monkeys were subject to a period of 24 hours of water deprivation. Dominance hierarchies were determined by the order in which the animals drank the water after this period of prolonged deprivation, but again a stable dominance order was demonstrated.

Marsden (1968) pointed to the fact that the behaviour of young rhesus monkeys can be considerably altered by the particular dominance–behaviour patterns of their mothers. It is too early to say yet whether some permanent change is produced by change of mother status, and obviously much more work needs to be carried out on this problem, but certainly the offspring reflected the change of the mother's position within the hierarchy, and indeed in many ways contributed to it.

The relationship of social dominance to the conditions of early rearing in primates has been investigated (Angermeier *et al.*, 1967). The heavier the animals, the higher their position in the dominance hierarchy, but the differential effects of their early rearing conditions appeared to play little part in determining the adults' ultimate dominance stature. They found, however, that one of the most important effects stemmed from a home cage influence. When two monkeys have had an opportunity to establish a cage as their home environment, they band together as a group against an additional monkey coming from another environment to join them.

In another study Angermeier *et al.* (1967) showed that the previously achieved dominance status of a monkey was an important factor influencing the formation of a new dominance hierarchy. Later (Angermeier and Phelps, 1967), it was reported again that previously achieved dominance status and previous experience in one dominance hierarchy exerts a profound effect on the status occupied in the next hierarchy. No biochemical differences could, however, be detected over the range of this investigation between the most and the least dominant monkeys. If primates are given complex discrimination tasks to perform, the behaviour of the less dominant animals appears to be affected by the condition of performing whilst other animals are present. If these less dominant animals are moved into comparative isolation, then they do indeed show an improvement in performance and seem better able to carry out the task which has been set them. Social status certainly appears to exert an important influence on the level at which an animal can perform these tasks (Angermeier *et al.*, 1967, 1968).

It is possible to argue that in the past greater theoretical significance has been attributed to the studies of dominance than perhaps they deserve. Some animals are obviously heavier or stronger than others. In any competitive situation the heavier and the stronger animals would be expected to win out. Dominance would appear to follow as a natural corollary of these physical differences between one animal and another. Given that competition is occurring, these differences in strength and weight would be expected to act back reflexively upon individuals to increase the weight and strength of animals succeeding in competition and to decrease comparatively the weight and strength of those animals losing in competition. Competition for the available food supply would be an obvious case in point, but there are many other factors besides this which would lead to an accentuation of the differences between the weak and the strong. The weak are more likely to fall prey to infection and the various diseases which affect the particular animal species. They are more likely to be injured by aggressive encounters; they are less likely to respond generally to stressful situations in a way which will leave them without some deleterious effect. All these factors, besides that of food competition, can act to magnify differences in the strength and the capacity for survival between one animal and another.

Once, however, dominance relationships have been established, they could well be maintained by a process of social conditioning. The infliction of pain in a natural or an experimental situation is something which promotes very rapid conditioning. An animal losing an aggressive encounter with a stronger and heavier animal is unlikely to escape without some punishment. The defeated animal rapidly learns, if possible, to avoid the previous adversary. Because of the social arrangements for living, this may not always be possible, and then the defeated animal must learn any pattern of behaviour that is effective in diminishing the attention of the previous adversary. The animal will also have learned those signs which indicate a potential aggression in the adversary, and these will now serve as warning signals to call out behaviour designed to placate the aggressor.

One of the most striking observations about dominance behaviour (to be examined fully in the next section) is that an animal, having once established a home territory, may fight to defend this particular area, and not only will become aggressive towards a newcomer but by virtue of previous tenancy will become dominant over him. Whilst the view expressed here is that dominance arises as a natural corollary of the fact that some animals are stronger than others, and that any factor which leads to an increase in aggression, whether it be prior training or injection of androgens, will also be associated with an increase in dominance, nonetheless the fact that animals fight to defend a territory and become more aggressive and dominant in this pursuit is a finding which needs further explanation.

Territoriality

Eliot Howard (1920) is generally credited with the development of the concept of territoriality. He was an ornithologist who during his work noticed that the males of many species of song bird early in spring restricted their particular singing activities to individual trees, shrubs or fence posts. Other males of the same species rarely appeared within the region of the singing male, thus Howard concluded that males of the same species divide their breeding ground into territories. Each one of those territories, he supposed, is occupied by a single male. At a later stage, females attracted by the song of the male join him, and also become active within this particular territorial area. Animals may also occupy a home range in addition to territory. This generally refers to any part of the environment in which an animal spends its time; the territory is distinguished from this as an area which is specifically defended against intrusion, and a region which the animal will fight to preserve. Hediger (1955) talks about the individual distance which one animal maintains between itself and another and, whilst this concept of individual distance is closely related to that of individual territory, it does allow for the possibility that an animal may defend a region around it or its family group which may vary with the movement of the individual – a mobile territory, so to speak, which accompanies the animal.

Throughout the literature there are very many examples of territorial behaviour in a wide variety of species. It is possible to give a few examples of the type of behaviour which has been reported. Pukowski (1933), for example, described the behaviour of the burying beetle (*Necrophorus* spp). These animals eat carrion which is lying on the ground. Both sexes search for it and, having found it, begin to bury it. One breeding pair of these beetles is found at a single corpse. Each single unmated male lays claim to a piece of carrion. He displays near this carrion, taking up position on a stem of grass or a nearby stone. The head is lowered and the abdomen raised, exposing the scent glands. The female, apparently attracted by this, flies in to join the male if his display has been successful. Mated animals will fight off intruders and ensure that finally only one pair is left in possession of the brood.

Fiddler crabs also show a marked territorial adaptation (Crane, 1941). Males and females of many species construct for themselves individual burrows, generally in the mud or sand of the intertidal zone. Males and females show a very strong attachment to the sites of the burrows, which have become a home and a place of shelter to these animals in more than the physical sense. Most of their time is spent around the region of the burrow, and at the slightest alarm the crab dashes back again to its burrow. Animals which have been snared (Pearse, 1914) and removed a distance from their burrow are set upon by every crab whose burrow they approach. Each territory owner is doing his or her best to defend this small area against the influence of intruders.

Many fish, either singly or in groups, maintain a territory which they are prepared to defend. Möhres (1957), for example, has shown that the electrical discharge activity of mormyrid fish is related to their activity of guarding their territory. When one individual had been allowed to establish itself clearly as a resident within a particular area, the introduction of another animal induced an intense discharge from its electrical organs. There was a marked change in the strength and rhythm of the discharge. The animals began synchronizing their discharge with one another, thus indicating clearly that the change was one of responsiveness to each other. The electrical discharge then followed a well-marked path, in which the ritual was carried through by both animals.

Eibl-Eibesfeldt (1960) reports an interesting case of territorial behaviour directly linked to feeding habits in fish. Anemone-fish of the Maldive and Nicobar islands live successfully amongst the outstretched tentacles of sea-anemones. They appear to have developed an immunity to the stinging action of the hematocysts by which the sea-anemone captures its prey. A substance produced by the skin of the anemone-fish inhibits the trigger action of the small harpoons which the anemone fires into its prey. The anemone-fish appears to adopt one or more anemones as its own. They become its individual territory from which other similar fish are excluded by a variety of threats and, if necessary, intense fighting. The fish appears to have secured its food supply for itself, and in so doing the anemone has literally become that animal's territory which it is prepared to defend.

Perhaps the most familiar example of territoriality is the behaviour of the domestic dog. Territory seems to be established largely by the scent of each particular animal. When two male dogs meet, they greet one another by a sniffing ceremony, by which they become aware of each other's particular smell. It seems likely that urination at street corners or any convenient object serves to communicate a male's particular claim to a certain area, or acts as a sign that a particular male is in the region and thus as a warning to other animals which may be around.

It has been widely suggested that mankind is particularly under the influence of the need for territorial possession (Ardrey, 1963). Territoriality has been used in sociological terms as a means by which communities may be analysed, and investigations have been made of the territorial behaviour of particular human groups. Altman and Haythorn (1967) isolated men from the rest of the community in groups of two. A gradual increase in territorial behaviour was discovered, which increased the longer the men were kept together. This followed a fairly orderly sequence. Each man gradually began staking out a geographical area as his own private territory from which the other was almost entirely excluded. Personal objects assumed significance as markers of this territory. The more mobile, less personal objects became incorporated within the territorial area only towards the latter stages of the experiment. The less compatible the people, the greater the strength of this territorial development.

There are many examples of territorial behaviour at the human level which could be quoted. Whilst the presence of certain forms of territorialism may be acknowledged, the question of its ultimate significance with regard to other forms of behaviour is a much broader question, relating to the problem as to why territory is maintained in the first place.

Reasons for territorial defence

There are several ways in which the lives of animals are affected by the fact that they hold territory.

First of all, the holding of territory is of great importance in the reproductive process. The territory forms an area in which reproductive pairs can indulge in courtship behaviour, free from the interference of other members of the same species. It is an area in which the young can be reared in a relatively peaceful fashion and where interference with the breeding process is reduced to a minimum. The defence of this area from intrusion has been proposed as an important function of territorial possession. It seems likely that it is particularly important to preserve an area which is free from interference from animals of the same species as the breeding pair. This is necessary in order to counteract the gregarious attraction which animals have for one another, the attraction which other males would have for the already mated female and the interference which other animals might effect with the young. One function of territory is to preserve the integrity of the animal family and, in so doing, to ward off interference from other individuals.

Another approach related to the idea of freedom from interruption concerns the nature of stimulation (Pontius, 1967). This view supposes that the animals, in maintaining a territory, are in fact looking for a peaceful place which will not act to overstimulate their nervous systems. In living within a particular area, much of the stimulation arising through novelty within the environment is reduced, simply because the animal becomes very familiar with its surroundings. In selecting one particular mate the social stimulation involved in selecting a mate again becomes severely reduced. The animal grows familiar with only one mate, and there again the necessary environmental stimulation is reduced. Pontius suggests that overstimulation of the limbic system is thus reduced. It is also possible to suggest that interference with the reproductive process by overstimulation of the adrenal-cortical axis is avoided. There is, however, a consideration which lies beyond that of an adaptation to stimulus levels; this is the fact that, with a reduction of stimulus levels through familiarity, an increased sensitivity to any disturbing aspect of the environment goes hand in hand. If the familiar environment provides little in the way of stimulation, then any intrusion upon this environment will by contrast be dramatically important and highly noticeable. It is possible, therefore, that the holding and maintaining of a territory leads to a marked increase in awareness and sensitivity to danger from intrusion, and that territory in this sense acts to sharpen the parent animal's response to danger, so that the young by this protection ultimately stand a much greater chance of reaching maturity.

A third view of territorial behaviour is that concerned essentially with the regulation of animal population numbers through bringing about dispersal (Lack, 1954; Wynne-Edwards, 1962). Success in claiming a territorial area usually ensures an adequate food supply for both the parents and the young. Defeat in the rivalry for a territory in many cases leads the animal to leave that particular vicinity altogether, in order to seek success elsewhere. The defence of territory which has been previously claimed and the lack of willingness to share it with others lead to dispersion of an animal population over a much more extensive area than would previously have been occupied, ensuring that limited food resources are not over-utilized and that ultimately survival of the species is enhanced, because of the more equal distribution of food and because of the enhanced chances for survival in widely distributed species members.

It is necessary to point to two more additional factors which appear to be important with regard to the tendency of animals to adopt and maintain a particular territorial area. The first concerns those predators which feed upon particular animal species. It has been pointed out (Tinbergen, 1957) that spacing of animals makes it difficult for predators to extinguish all of the members of a population within a given area. There are, however, far more insidious predators than the ones which Tinbergen was thinking of. These are the disease processes, the epidemics, the viruses which may on occasion destroy whole populations. Disease processes of this kind must be of enormous significance with regard to adaptive evolution and the complete survival of animal species. Socially living groups have been destroyed by disease processes because they could so easily assume epidemic proportions. Isolated animals and groups of animals separated from the rest would stand a greater chance of avoiding this, and thus there would be a premium on the capacity to disperse away from members of the same species. In addition to the capacity to secure a reproductive area and an adequate food supply, the avoidance of predation of whatever kind would appear to be an important factor in animal distribution.

Another factor leading an animal to adopt a territory may be that of exposure learning. We know that animals, particularly birds, learn to choose between familiar and unfamiliar aspects of the environment at a very early age. This usually applies to a preference for familiar rather than unfamiliar objects in imprinting, but the animal also learns the characteristics of its home cage, which it prefers over other areas. In addition, chicks seem to show imprinting with regard to the food which they are fed. It seems reasonable, therefore, to suppose that animals adopt a particular area as their own through a process of exposure learning, and that perhaps they are guided to some extent in their choice of area by the nature of their own early experience. However, be that as it may, many animals adopt an area as their own; presumably this area has now become familiar to them and they are prepared to defend it against intrusion from any object which seems foreign to it.

Leadership

The nature of social dominance existing within an animal community obviously bears a close relationship to the leadership of the group. Any community of socially living animals capable of responding as a group must be under the influences of forces acting to integrate the movements and processes of the group. If each animal acted as the controller of its own behaviour, a chaotic lack of integration would appear, which would reflect the random organization of the community. It is possible, therefore, that the purpose of social-status relationships is to integrate socially living animals into a finely poised and coordinated system which ensures that group activity rather than individual chaos governs the behaviour of the social unit. This is as true of the family unit as it is of the larger group. Social dominance allows for the emergence of one obvious leader, whatever the initial organization and grouping of the animals. If the existing structure should through any reason become disorganized, then social dominance allows the units to be reassembled and the leadership pattern to be once again established. Leadership entails more than an appraisal of the relationship of the group to external events and organizations. It carries with it the responsibility for control and function within the group.

The dominant animal is usually the strongest and the heaviest of the group. This may mean that this animal is probably the bravest and the one with the least to fear from attack. This animal is probably also the least fearful. These may be the characteristics which are important in occupying the predominant position. This animal, because of its physical characteristics, is also most likely to predominate in the defence of a home range or territory and to be successful against intruders and strangers to the group. Other animals more fearful within the group, by aligning themselves with this dominant animal, also achieve protection and, in so doing, strengthen the group integration and its capacity for defence. The question of leadership is not, however, as straightforward as it may at first seem. It is possible that the leader acts as an integrator of the information which he receives. It is possible that for much of the time he coordinates the group activities, but it is not clear how far decisions are the prerogative of the leader or how much the decision to act is that of the group which is served by the leader. Certainly each leader of an animal group would be expected to be highly sensitive to the actions and responses of the rest of the group, just as a cabinet minister receives advice and briefing from members of the civil service. Actions from the rest of the group could both represent group decisions in their own right, as well as the information on which the leader is required to make his decisions. The nature of leadership is a very complicated one. In most cases leadership is, however, closely tied to the processes of social dominance.

Summary

There are three major areas which relate to the exercise of status and hierarchy. These are dominance, territoriality and leadership.

Dominance is a well-attested fact in animal behaviour. One animal exerts superiority over the other in a variety of different situations. This relationship may be determined at the first encounter. Animals may fight or express hostility towards one another until ultimately one emerges the victor. After this, the status relationship remains as relatively unchanging, each animal occupying a characteristic position with respect to the other. The features which appear to determine social dominance are physical strength, lack of fear, the opportunity to fight on home ground, seniority, physical health, fatigue, concentration of sex hormones and individual training.

One of the striking observations with regard to dominance relationships is that an animal wins far more encounters and becomes much more dominant on its home ground. The holding and maintenance of a territorial area is one of the key factors in status and hierarchical relationships. Not all animals defend territory, but there are many which do. It has even been suggested that territorial adaptations are important with respect to human functions.

Territory holding is important in the reproductive process, both in attracting a prospective mate to the courtship area, and in providing space in which to rear the young where interference is reduced to a minimum. The spacing of individuals may be relevant to the problem of predation by other animals, as well as to the problem of infectious diseases. It may also be important with regard to available food reserves. It is suggested that animals adopt a particular area as their own through a process of exposure learning. They are guided in their own choice of area by the nature of their early experience. Animals visiting the area are not familiar to the animal defending it, whereas the rest of the area is totally familiar. The defender has a basis for differentiation and strange intruders are evicted from the area.

Status relationships may be expressed through the leadership of the group, which bears a close relationship to dominance and the maintenance of territory.

Group Behaviour

The behaviour shown by animals when they aggregate together in groups poses many interesting problems concerned with group management, coordination, communication, control and organization. There are many studies of group behaviour ranging over a wide variety of species. These studies in the main owe an allegiance to the tradition established by the early naturalists. They could be said to form the foundation of a sociology of animal behaviour, although in fact they are often limited by a failure to penetrate beyond the descriptive level and a failure to advance beyond an account of individual behaviour patterns within the social context.

The experimental approach has been applied successfully to the study of social behaviour but in respect of group structure and organization, faces the problem of artificiality. An experimental group may behave in a different fashion from a natural group, but this is not as serious an objection as it might at first seem, because cross-referencing can take place between natural and experimental groups and knowledge about behaviour gained under controlled conditions can be substantiated by natural observation. Experimental techniques are used to analyse the nature of animal society, and in time could form the basis of a vigorous discipline rooted in animal behaviour, which would take the form of a truly experimental sociology, concerning itself not with the behaviour of the individual but with the organization of the society, the family, the tribe and the race.

It has been suggested that the study of animal society has unique importance in that it allows man to hold up a mirror to his own society. It is believed here, however, that the usefulness of this comparison is very much exaggerated. The differences between the societies of man and animals appear considerably to outweigh the similarities. It is suggested in a later chapter that statements about the social function of man based upon animal behaviour should of necessity be substantiated by observation at the human level. Nevertheless, the study of the organization of animal society does have advantages for a variety of specialist purposes, and in addition to its intrinsic interest there may be occasions when it could stand in place of investigation at the human level.

Types of social organization

Animal societies are highly diverse in their nature. Each may be composed of a number of different social arrangements. Each social unit may interact with other units, and parts of one may overlap with parts of another. Smaller units may be encompassed by larger units. However, it is possible to distinguish five main types of social organization.

First, there is the reproductive pair, a group distinguished by the formation of the pair bond. The sexual relationship may be brief, animals coming together for only a short time, and then with a succession of partners; or the pair bond may become strongly established, the animals persisting in a more or less stable union. Pairs of

animals establish many different kinds of relationship
between themselves, in addition to the reproductive one.
These have been described as dyadic encounters. Dominance relationships may be of this kind, but tend to be a
disruptive rather than a cohesive influence and do not in
themselves lead to a close association of the animals concerned.

Secondly, the family group represents an arrangement
resulting from the extension of the group formed by the
reproductive pair. This group, as its name suggests, consists of those animals which are united because they belong
to the same family. Some family groups tend to persist after
the young have reached maturity, and may extend over
several generations, but as a general rule the period of
maturity signals their dissolution, and the young separate
themselves from the parents or the parents drive away the
young.

Thirdly, the short-term larger group may consist of a
temporary, loosely integrated collection of individuals who
associate together for a while and who then separate off
from the group, either to go their own way as individuals
or to join other groups.

Fourthly, long-term groups, on the other hand, have a
much more closely defined structure. It is only over time
that the relationship of each member with the rest of the
group becomes clearly established, and it is in the large
permanent group that the system of rules for social conduct
becomes most firmly established.

Finally, society itself is composed of many or all of these
different types of group. It is at this level that the large-
scale functions of populations become important in their
relationships with the rest of the environment.

Some of the most striking of the social organizations of
animal communities are to be found amongst inverte-

brates. These are very numerous, and it is possible to describe here only what are perhaps the best-developed and most studied communities. Perhaps the best-known community of all is that of the beehive. A normal honeybee colony, during the summer at least, is composed first of the queen. She is specialized principally for egg laying. Her task is to maintain the reproductive facility of the community. Secondly, there may be several hundred drones. These fertilize the virgin queens but apparently serve no other overt function. Thirdly, the workers, whilst morphologically indistinct from the queen, bear the sole responsibility for the activities other than the reproductive process itself within the hive. Workers live for four to six weeks. There may be several thousand workers in the hive. In addition to these adult animals, there is the brood. These are housed in the wax combs which form the physical furniture of the hive.

The queen is frequently attended by a small number of worker bees. These congregate around her when she is laying eggs, or when she is stationary on the comb. Bees take little interest in the queen during winter (Allen, 1957), but during the summer the attendants examine her with their antenae, lick her and presumably feed her on glandular secretions.

Whilst a small band of animals at any one time devote their attention to the queen, there are many other duties in the hive which need to be performed. One of the first functions which falls to the young worker is that of cleaning cells. Rösch (1925) observed that a newly emerged bee begs food from older bees and begins cleaning, not only itself, but also cells which have formerly contained the brood.

Rösch marked particular bees and observed that they performed specialized tasks according to their age. The results suggested that individuals were not specialized for the performance of a particular task, but each bee performed a variety of tasks, the nature of the task tending to change as the bee grew older. The first tasks are those of cleaning and maintaining the hive. The cells are cleaned by the young bees before the queen lays her eggs in them. Any cocoon debris is removed with the mandibles, and remaining larval excreta is covered and insulated with a thin layer of wax. The bees fly from the hive carrying debris of various kinds. Rösch reported that old animals, already familiar with the terrain, are chiefly responsible for the removal of the debris from the hive and have already undertaken orientation flights. Worker bees clean other bees. A bee wanting to be cleaned performs a grooming dance, during which it shakes its body and stamps its legs.

One of the most important activities within the hive is that of nursing. The nurse bees provide food from their hypopharyngeal glands and mandibular glands which they give to the larvae. Quite young members of the hive have been shown to feed larvae. It is not clear exactly how workers know when to feed each member of the brood. It has been reported, however (Gontarski, 1953), not only that the larvae are given an excess of brood-food if it is produced in large quantities, but also that when there is insufficient food, there is no equality of distribution. Some larvae are adequately fed, the rest are ignored and left to die. Queen larvae are fed at the expense of worker larvae (Simpson, 1957). Some animals perform what may be guard duties, standing at the entrance of the hives and rapidly examining incoming animals with their antennae. Some animals seem to do this before becoming foragers. It is not always clear that these animals have been set to perform these tasks, and this raises serious questions as to what guard duty really means.

The problems of explaining how this social organization comes about are enormous. Lindauer (1952) suggested that in wandering around or patrolling the colony the house bees, in constantly inspecting the cells and brood, gather for themselves information about the jobs which need doing. Each bee inspects and regulates the function of the hive by its own labour. Nixon and Ribbands (1952) allowed foragers to collect small quantities of sugar syrup containing radioactive phosphorous. This was soon widely disseminated throughout the members of the colony. As a result of this evidence Ribbands suggested that food transfer from one animal to another makes members aware of changes within the colony. The exchange of nourishment could certainly act as an integrating mechanism. It is possible that chemical communication between one animal and another could also be established by this means.

It could, on the other hand, be suggested that feeding is just another form of behaviour and that the principles of community organization are not necessarily embodied in it. Similarly, the well-known dance behaviour may reflect rather than be the means of organization within the community. Many suggestions have been put forward as to the purpose of these dances. Milum (1958) supposed that they alerted the colony to activity; Wittekindt and Wittekindt (1960) suggested that they are recruiting dances. Whatever the nature of dancing behaviour, it is clear that the problem of the nature of social organization is one which seems to be over and above these individual means of expression.

Wallis (1965) has described the social organization of the ant colony. In some cases ants may show physical changes which enable them to be specialized for particular tasks. Most commonly, however, individuals seem to prefer to perform one particular task, but may on occasion perform others. The queen in each colony performs the major task of egg laying, and possibly she is partly responsible for the care of the young. Queens may on occasion, particularly during the founding of a colony, show many of the behaviour patterns of the worker. They may forage, construct simple nests and show aggressive behaviour. Males, on the other hand, have not been reported to show the behaviour necessary for the tasks which the workers perform. Their function is reserved principally to mating. The workers perform the tasks of foraging, fighting, nest building and care of the brood. Some ants have the facility to perform what appears to be slave-making activity. Whether this is an example of cross-species cooperation or presupposes the type of organization one is accustomed to think of in human terms is not entirely clear. It is reported that both the blood-red slave-maker (*Formica sanguinea*) and the Amazon ant (*Polyergus rufescens*) both make slaves of *Formica fusca*.

Termites present a picture of a complex social organization (Brian, 1957; Weesner, 1960). Unlike ants, both sexes play important roles throughout the structure of the colony, and in the termite colony there may be many immature individuals, either in a stage of progress towards the mature insect form or in a state of arrested development. The reproductive pair are the founders of the colony. These are a pair of mature insects which have flown from another colony. On their first meeting outside of the nest, they shed their wings along the basal suture. Apparently the stimulus of contact with a member of the opposite sex is sufficient stimulus to bring this about. This reproductive pair form a community which shortly gives rise to other reproductive pairs, when the first reproductive pair have exhausted their potential. The development of nymphs may be accelerated to bring this about, or already existing adult forms, awaiting the opportunity to swarm, may form reproductive pairs. When the task of reproduction has been designated to certain individuals, their wings are forcibly removed by other members of the colony.

The animals which are not required to reproduce fall into four categories: larvae, nymphs, soldiers and workers. Larvae and nymphs appear to play little part in the control of the activities of the colony. Nymphs in some species are, however, able to feed themselves, and in fact take an active part in the trophic food circulation.

The soldier caste develops from the larvae after two months. At first a 'white soldier' is produced. This animal is feeble and inactive. After another month it turns into the fully formed, sclerotized soldier. Soldiers, naturally enough, play a large part in the defence of the colony. Some species develop a special head capsule which is used to block tunnel entrances. Some soldiers have highly developed menacing mandibles which are used as sharp-edged shearing weapons. Mandibles may also have teeth or serrations on them, which enable the termite to maintain a firm grip on the adversary. The soldiers have no eyes but nonetheless they guard entrance holes and protect foraging columns of workers (Harris and Sands, 1965).

The workers are largely responsible for gathering food. Food is dispersed throughout the colony by a process of social rumination. The food, which is usually wood and woody structures, is not completely digested in its passage through the alimentary system of one animal, and it may be regurgitated or defecated in order that other termites may redigest it. The workers emerge from the nest in columns guarded on each side by soldiers, in order to forage for food. They build special tunnels and platforms between the nest and the food in order to be able to do this.

The workers of many species maintain what is known as a 'fungus comb'. They deposit faecal or chewed material in a certain place to build up a comb, and on this the mycelium of fungi is grown. The comb is then eaten and replaced by fresh material (Kalsoven, 1936). The workers also perform the highly important functions of supplying the king and queen as well as the larvae with nourishment, in the form of saliva enriched with the products of digestion. Workers also clean and tend them.

It is clear that highly organized societies exist at the invertebrate level. It would be wrong, however, to suggest that all invertebrate social life is organized in this complex way. Nevertheless these studies do point to the need for caution in the consideration of the evolutionary view that animal societies show a progressive advance towards that ultimately displayed by humans.

Social grouping has frequently been discussed in rodents (Barnett, 1963; Archer, 1969), in birds (Crook, 1961; Moynihan and Hall, 1953; Morris, 1956), in rabbits (Southern, 1940), in cats (Leyhausen, 1965) and in dogs (Fuller and DuBuis, 1962). There are many investigations of the structure of primate groups (Carpenter, 1940; Southwick, 1962; De Vore, 1963; Washburn, Jay and Lancaster, 1965; Hall, 1965; Gartlan, 1966).

Lawick-Goodall (1968) has described the nature of group structure amongst free-living chimpanzees. The association between the mother and the young appears to be the most stable of the groups comprising the social organization. What appears to be an organized group containing older animals may be nothing more than a temporary association of a few individuals. Dominance was clearly observed between one animal and another, but it is not clear exactly what role the dominance of one animal over another has in the organization of the social structure.

It appears that leadership is to some extent taken by the most dominant animal of any particular group, but the leader may not always occupy the position at the front of the group. When animals are aggregated together in large groups, such as the bachelor groups, an obvious leader is by no means apparent. Groups of chimpanzees range through quite a large area. Groups coming into contact with one another on their wonderings do so without conflict, and groups may unite from time to time without any overt show of aggression. When groups came into contact there was, however, frequently a show of excited behaviour. The males drummed on tree-trunks with their feet and shook or dragged branches, slapping and stamping on the ground.

The relationships between infant animals were characterized by a great deal of play behaviour. Young infants spend most time in play, but juveniles also play a good deal. Lawick-Goodall observed infant behaviour on several occasions following the death of the mother. This circumstance formed a striking natural experiment. The infants grew listless and showed a low frequency of play. One infant became emaciated and socially abnormal. The death of the mother in these few recorded instances appeared to exert a profound effect upon the infants. The infants were not outcast or ignored by the society. They had frequent contact with members of the group, and in two recorded instances the infants surviving the death of the mother were adopted by an older female sibling. This investigation also showed that chimpanzees kill and eat fairly large mammals and that they form primitive tools out of a variety of natural objects.

The association between the mother and the young appears to be the most stable of the groups composing the social organization. This study points to a need for caution in describing group behaviour, because what appears to be an organized group of older animals may be nothing more than a temporary association of a few individuals. It may persist no longer than a few hours or days, and indeed mature animals frequently move about on their own. Highly elaborated social control is not allowed for in characteristic temporary organizations of this kind.

The investigation of group behaviour under wild or semi-natural conditions is beset with many difficulties, against the background of which the results have to be evaluated. First, the investigator introduces uncertainty into the situation, in that his presence can influence the group structure and the nature of the behaviour he observes. This may be particularly true of ecological investigations, where the observer by his presence may drive fearful animals away from the area in which they had formerly been living. Lawick-Goodall (1968) reports that the chimpanzees she was studying at first ran away from her. To overcome this difficulty, artificial feeding was undertaken, but this again is a new element in the situation. The observer's friendship with the animals under observation introduces another new element into the consideration of group structure.

Secondly, in most cases naturalistic observations do not permit a finer analysis of causal relations. The investigations are largely descriptive and by their nature designed to catalogue the types of behaviour which individuals show; but is this sufficient to answer questions about the nature of social organizations and control?

Causal factors are also difficult to isolate in this kind of study, because any one event is accompanied by many others. Which of the preceding events in the social life of the community is the one which led to the behaviour now under observation? The work of basic description provides for the enumeration of particular behaviour patterns. We must take it on trust that these are as separate and discrete from one another as we have been led to believe. This work, however, is far from satisfactory in providing a deeper understanding of social behaviour, and investigations are now needed which penetrate beyond first-order observation into the causal structure of the communication system and the chain of command within the group.

Communication

One of the important factors in the establishment of cohesion between group members is the communication which exists between them. Communication is a process which entails a chain of events leading from the behaviour of one organism to the sensory and interpretative systems of others. Communication can only be specified as having taken place if it is possible to observe a change of state in the receiving organism. To suppose that communication is achieved only in terms of a few simple learned habits or alternatively in terms of a few inbuilt responses, is to ignore the complexity and subtlety of most forms of animal communication. To suppose that animals use at best a reflex system, by which several stereotyped response patterns are put into operation, seems to be an oversimplification.

It is frequently supposed, for example, that man has the monopoly of language and that this is the feature which uniquely distinguishes him. This view not only elevates human performance but also debases the nature of animal communication in a way which does not truly fit the facts. Bird song is a form of complex vocal communication, which in many ways could provide a parallel for the development of speech in man. Birds not only show a typical species song but also give rise to different groups or familial dialects (Thorpe, 1961). Dolphins also have an extensive vocal range (Kellog, Kohler and Morris, 1953). They give rise to different bird-like whistles, which appear in different melodies and patterns, as well as clicks which vary considerably in their tempo. Animals which find themselves in distress emit calls of two rather different whistles. These are repeated again and again. When one animal emits this particular call, all the other animals remain silent for a short while and then begin a search for the individual in distress. This animal may then be pushed upwards by the others to the surface and the helpers carry out whistle exchanges with the distressed animal (Lilly and Miller, 1961). Sound patterns of this kind are used extensively in social communication and, if a radio communication link is provided between two previously isolated animals, then they use it to communicate in tight bursts of sound (Long and Smith, 1965). By this means it is possible to establish something of the meaning of the whistles used in communication. One type of whistle appears to act as a call sign to establish subsequent claim to the radio link. The other calls correlate closely with the vocalizations of the other animal. They are very flexible, and the animals appear to use them in what can only be described as a conversation.

Not all animal communication need be established by vocal means. In many cases this purpose can be served by musculature other than that of the vocal apparatus.

In most cases there are plenty of means by which signals may be provided and from which a language can be elaborated, but the system for receiving and interpreting information, as well as the means of sending it, must be considered.

The term semiotics has been proposed by Margaret

Mead to cover the study of the human use of signs and the way in which they are organized in transactional systems. Sebeok (1965) suggested that zoosemiotics is a term which should be used to identify the rapidly expanding study of animal communication. This can be regarded as a dominant theme in the study of ethology and comparative psychology; it also forms a principal axis on which the study of social behaviour in animals rests.

In essence zoosemiotics attempts to apply concepts derived from the study of communication and information theory to the interaction between one animal and another. It parallels an important movement in the study of human communication (Broadbent, 1956). It argues that the position with which the investigator of animal behaviour is faced is that of the antiquarian trying to unravel the code of a language appearing in script, but a language which no contemporary understands. Similarly, the investigator of animal behaviour tries to unravel the language by which one animal communicates with another. In order to understand communication there are several stages which must be rigorously investigated.

First of all, messages are sent through a particular channel. It is important to identify the channel and, if other channels are being used, to specify exactly which of these is contributing to the communication. Channels may supplement one another in transmitting information, particularly over long sequences, and may operate to provide communication which may at first glance appear irrelevant to the particular message.

Having established the channel of communication the next task is to investigate the means by which the animal sends the message. In some cases, as in echo location in bats (Griffen, 1958), and in the measurement of the distortion of the electrical field emitted by the fish *Gymnarchus*

(Lissmann, 1958), the message sent may be reflexively re-received by the animal itself and used, for example, in orientation. Usually, however, the message is directed, not back to the individual, but towards other individuals. The organism may select a message out of a code of behaviour it possesses in common with the species, or it may communicate by particular signs which it has learned. In each case it could be argued that it is using a language to transmit information. It is not sufficient, however, that the organism should do this for it to be said that communication has been established. It is essential that the message should have been received. This is not a simple matter, because the message is framed within a particular code which may be unique to a particular species. It could be given as a genetic endowment of the organism, or it may be acquired through the experiences of the organisms, one with the other. In order that the message can be received, an organism must understand the message within the meaning of its own code and, in 'understanding' it, and in responding to it, must act as a decoder to the message. To the observer, however, the recipient organism can only indicate that it has received the message by some change of its own behaviour. The study of semiotics therefore includes this rather complex chain from the sender to the recipient, then ultimately to the observer.

Whilst the observer may determine that a change of state has occurred in the recipient organism, this is only sufficient for him to establish that a communication could have taken place. It does not allow him to establish what the communication was. His task is then to try to break the code, so that he too may understand the message which has been transmitted. Goodall (1965), for example, came some way to penetrating the communication barrier existing between chimpanzees and human beings. She describes how after a time she was greeted as another chimpanzee. Communication has been studied in this way as an attempt to understand animal language.

It is possible to gain evidence about the nature of communication in animals reared in isolation. Will they show the typical patterns of behaviour used as the means to transmit information? Will they show the capacity to understand the communications of other animals if during infancy they have been denied the opportunity to acquire them themselves?

Chaffinch song has been extensively studied by Thorpe (1954, 1958). Chaffinches which have been isolated and hand-reared develop a very simple type of song. This is of approximately the right length and has the customary number of notes, but it is a song which is undistinguished. It is not divided into phrases and it lacks the terminal flourish. If the opportunity for listening to other animals does not exist, then a song is developed which lacks the nuance and elaboration of the social-living animal. If nestlings are brought up in groups in isolation from adult birds, they sing songs which are more elaborate than those of totally isolated birds. The song patterns show similarity between one member of a group and another, but they may differ widely between the groups. Song formation reaches its peak in free-living socially reared birds. These animals give rise to the fully elaborated song.

In both the blackbird (Messmer *et al.*, 1956) and in the meadowlark (Lanyan, 1957), individuals reared in isolation also produce only a simple song. The full song only occurs when the animals have been reared socially.

The possibility remains that learning may play a part in the generation of simple songs, although the animals are isolated from their companions. They hear their own vocal productions, and learning could assist in control of singing as the development of a feedback loop, as well as in the association of certain sounds made by the animal with particular emotional states. Deafening, however, does not appear substantially to affect the simple form of song. Messmer *et al.* (1956) report only minor differences in tonal quality in deafened animals, and Kanishi (1963) obtained similar results with chickens.

It appears from this work that some birds, although reared in isolation, are capable of producing a simple song, but that the development of this song is enhanced and elaborated by the presence of companions. Isolated groups develop song patterns of their own which differ significantly from those of other similarly isolated groups. It is possible to suggest that the development of song represents something more than the imitation of the song of one animal by another. It represents the growth and the development of communication. Song is one of the important ways in which messages are conveyed from one animal to another. As in human language, there are regional differences and local dialects, so it could be suggested that there are local and regional influences upon the development of particular patterns of song. The suggestion arises that animals are provided with the basic apparatus for the expression of communication, but the language of the communication and the growth of the understanding of this language are something which comes about through the interaction with fellow members of the species. If the organism is denied the opportunity for this interaction, then it is no longer able to communicate appropriately, because it has failed to learn the correct code.

The suggestion that communication is established by a series of codes which are acquired in many cases during infancy need not be restricted in its application to the study of the development of song in birds. The work describing the social disadvantages and the deficit in social interaction in animals reared in isolation has been discussed previously. It has been assumed that animals reared in isolation are disturbed in various ways, and are thus led to show abnormal response to members of their own species. An alternative view, however, can be proposed, related to the effects of isolation on the development of communication. If an animal is denied the opportunity to interact with its fellows, then it fails to learn the language which is necessary to bring this about. This language need not be a vocal one. It could be a visual behaviour language; it could be a series of signs directed to any of the sensory modalities. If organisms learn these during early infancy, then an animal isolated during this time will not know the appropriate behaviour signs to display to its new companions, and will not know the appropriate vocalizations to make. It will not understand the signs made to it by the other organism. It will, in short, not be able to communicate with organisms of its own species. It will not know their code, with the result that communication will be impossible.

The disorders of behaviour shown by previously isolated animals in respect to members of their own species could reflect the behaviour of an animal, not necessarily suffering from some gross pathology of behaviour, but one which has failed to learn the language appropriate to communication with members of the same species.

Group factors and available space

One of the most rapidly developing areas in the study of the social behaviour of animals concerns the establishment of experimental populations. This is important in the study of population dynamics. Large groups of animals confined to small areas show certain abnormalities of behaviour. Reproduction may become difficult or impossible. Sexual behaviour is disrupted. In some cases copulation cannot take place because large numbers of males all attempt to mount one oestrous female. Indiscriminate mounting may take place at a later stage, and many rats show hypersexual as well as pansexual behaviour (Calhoun, 1962). Litter mortality increases sharply in overcrowded conditions as the result of either cannibalism or insufficient maternal care.

The most striking observation of these studies of overcrowded social groups of rodents concerns the nature of aggression. Overcrowded animals generally become much more aggressive than other animals, possibly because submissive animals cannot escape and also because the opportunity for aggression becomes increased with physical proximity. Reactive or vicious-circle aggression also increases for the same reasons. Severe fighting is commonly the outcome of overcrowded living conditions. Christian (1959) and Thiessen (1966) report aggression as the result of overcrowding in mice species.

Important observations on other species also suggest similar findings. Domestic hens, rabbits and cats all appear to show an increase in aggressive behaviour when placed in overcrowded conditions. Hutt and Vaizey (1966) added children to this list. They studied the effects of group density upon the behaviour of children in a playroom. Children became more aggressive as the group size increased, and this tendency was accentuated in brain-damaged children.

A discussion of the relevance to man of research findings, gathered in the first place from the study of animal behaviour, is undertaken later, and it is sufficient now to note that certainly in the case of the behaviour of children a fairly close parallel can be established between the response of infra-human animals on the one hand, and the behaviour of man on the other.

Problems of social organization

It has often been supposed (e.g. McDougall, 1908) that there are forces of instinct which bind animals together in groups. These, it is held, are responsible, not only for the fact that animals come together, but also for the high degree of cohesion and organization which many animal groups possess.

There are many objections to this view of social behaviour. First, instincts may be used not as explanatory principles but as imprecise constructs to fill a gap in our knowledge. Secondly, they may be used simply as another name for the behaviour which has been observed. For example, the explanation of gregariousness is not enhanced by the supposition of the gregarious instinct; this is just another name. Thirdly, much social behaviour can be shown to be learned and is not, properly speaking, instinctive at all. In addition Tinbergen (1951) denied the presence of a distinct social instinct. He supposed that there were no special activities, which could in their own right be called social, that were not part of some other 'instinct'.

It is equally possible to seek the origins of gregariousness and social cohesion in early learning. The experiences which animals have in belonging from earliest infancy to a family and social group are undoubtedly important in determining their social tendencies. It will be recalled, however, that in an earlier chapter the view was expressed that the analysis of behaviour into what is innate and what is acquired, and the attempt to ascribe behaviour to one category or the other, is an unproductive pursuit. It is not the intention here to attempt this kind of separation but to point to important factors in the development of social response.

The first factor to be considered is that of the gregarious tendency. Group-living animals isolated from their companions attempt to return to them usually as rapidly as they can. Certainly, young neonate chicks spend much of their time in close association with one another and will make every effort, almost immediately after hatching, to meet up with other companions in an incubator. This cannot be taken as unequivocal evidence of a social tendency, although this kind of behaviour undoubtedly serves the purpose of aggregating the animals together, because chicks are attracted to any moving object, particularly if that object is also making a sound. Moving inanimate objects are readily approached and followed.

If this argument can be applied to other species, it would have to be said that there are no necessary social tendencies, but that social aggregation comes about from the earliest stages by virtue of general mechanisms of locomotion and perception, which have the effect of bringing animals into close association. In the case of the infant mammal it finds itself in a litter with others, as part of the family in which the parents play a part in its socialization. This form of life then becomes part of its experience, and social tendencies are no doubt related to it. The young animal from its earliest days receives food, water, warmth and protection from the litter or the family group. Socialization is a process which, whilst it may begin after birth, need not be confined to the period immediately after birth. The advantages gained by infants from living in the family group may be largely those that an adult animal also obtains from living in a larger social group. Animals have previously learned

to distinguish their own species from others. Social rewards have come from members of the animal's own species
in the past. The means of communication between one
animal and another of the same species have already been
established. For these reasons and many more, it is not surprising that socially reared animals should show a tendency
to regain the companionship of others when they have been
separated from them. Animals may be expected to maintain themselves in social groups unless other factors in
development operate against this principle.

The next problem concerns the nature of the coordination which becomes established within group activities.
There is no doubt that animals show a great deal of social
coordination in their behaviour. There are several possi-
bilities as to how this is achieved.

Trophallaxis (Schneirla, 1952) appears to be one means
by which the activities of the insect community could be
coordinated. This is the process of social rumination. It is
conceivable that in the transfer of food from one organism
to another hormonal substances pass over, which have a
regulating function on the behaviour of the animal being
fed. Control over community behaviour would be established by chemical means. It is a long step, however, from
the demonstration of the presence of an ectohormone in
regurgitated food to a general theory of social regulation
within the community.

Whilst trophallaxis may be acceptable as a means appropriate to the control of social behaviour in insects, it cannot be used to explain the social organization of vertebrates. Even if trophallaxis may be regarded as a means by
which instructions are passed from one organism to another, we still do not know how such possible instructions
operate the chain of command or how these instructions

relate to the control or the reactive machinery of the social organization.

Another view, which hints at something of the organization necessary to achieve social coordination without actually stating how this comes about, is the supra-organism view expressed by Emerson (1952). This view supposes that a particular social community resembles the organization of an individual animal. A society is an organism of a particular kind. Emerson believes in the presence of fundamental units in evolution and, just as it is supposed that the multicellular organism evolved from unicellular units, so it is supposed that society has evolved from the aggregation of individuals. Natural selection operates upon the larger as well as the smaller unit to produce adaptations which are essentially analogous, in response to the same forces. Whilst this view has a lot to recommend it, it is not clear that the principles of organization of the individual are the same as those of the society, or that the forces acting upon the individual are either the same or affect it in the same way as those acting upon the society. In addition, this view still leaves open the problem of the chain of command within the society, and does not really explain the nature of the coordination of the activities between individuals.

It is possible that the appearance of coordination of activities may be achieved in animals which retain a close proximity to one another without any actual chain of command. If each animal responds in rather the same way to the same situation, then each will react as an individual, but social cohesion will give the appearance of group response. A group of antelopes, for example, may all remain vigilant during feeding, each one sensing danger and moving off at the presence of a predator. This will give the appearance of a group movement owing its origin to individual sensitivity. Group cohesion is the only necessary principle under these circumstances to give the appearance of group control.

Group cohesion argues for considerable sensitivity on the part of individual members of the group towards others. Cohesion may also be achieved through social facilitation and observational learning. These factors have been discussed previously with regard to motivation and social control. The behaviour of one animal could lower the threshold for the performance of that same activity in another. One animal may act as a model for another, or animals may simply learn that certain kinds of social behaviour in others, if carefully observed, bring themselves reward. For example, when one animal begins feeding, it may indicate the presence and location of food to others.

It is possible also that coordination may be achieved by a chain of command. The instructions are given by the individual or groups of individuals in authority. The rest of the group follow these instructions.

This is not the simplest explanation but it must at least be retained as a possibility, certainly with regard to the more complex of social activities.

Summary

This chapter has explored the many problems associated with group management, coordination, communication, control and organization. Examples are provided of the complexity of some organized animal societies. It is suggested that more attention should be paid to the problems of the control and organization of societies, than to the observation and description of individual behaviour patterns occurring within a social context. The problem of communication is discussed. It is suggested that animal communications take the form of a language and that their subtlety and complexity cannot be overlooked. The study of animal communication is described within the framework of zoosemiotics, which is the application of the concepts derived from the study of communication and information theory to the study of animal behaviour.

Social factors are discussed in relation to group pressure and to available space. Reproduction becomes difficult under these circumstances, but noteworthy also is a marked increase in aggression in dense populations.

Possible means for organizing and controlling group behaviour are considered. The idea of social instincts is regarded as unsatisfactory. Trophallaxis is seen as only one possible means of establishing control in insects, but it does not explain social organization amongst vertebrates. Another theory is that of the supra-organism. This view still leaves open the problems associated with the chain of command and does not explain the nature of social co-ordination. Integration of group activities by individual function with the addition of a cohesive principle is put forward as a possible means for coordinated activity, as well as control by direct command.

9   Animals and Man

There are many areas of medical research where experiments which can be carried out upon animals would be difficult, if not impossible, to conduct upon man. Genetic experiments are often of this type, as are many pharmacological and physiological investigations. The results obtained are used to stand in place of those which are not available from human investigations. They are used as substitutes and, if the biological systems of man and other animals resemble one another to such an extent that the results obtained from the former can stand in place of the results obtained from the latter in this type of investigation, might it also be the case that studies specifically concerned with the behaviour of one could also substitute for studies on the behaviour of the other? In other words, do the same broad general principles govern both the behaviour of man and other organisms? Can the results of experimental investigations of the behaviour of animals be used as a guide to the underlying determinants of the behaviour of man? It is the purpose of this chapter to try to discover something of these broad principles, and also to investigate the criteria by which evidence obtained from the behaviour of animals can be used as arguments about the root causes of human behaviour.

Historical viewpoints

In the early days of the study of animal behaviour one prevalent tendency was to regard animal behaviour as being determined by mental processes very similar to those possessed by human beings. The mentalist viewpoint attempts to interpret the lives and behaviour of animals in terms of human conduct. It is frankly anthropomorphic and is typified by the approach of many of the early naturalists.

Existence and self-perpetuation are not sufficient criteria for society. Awareness is the essential element. Something which McIver and Page (1949) call the psychical element must exist before there can be society, and the psychical element appears in this case to mean awareness. The study of social behaviour in animals is, however, a testing ground for these views. Few would deny that coherent societies exist amongst animal groups low in the phyletic scale. If the possession of minds makes society possible, then this must be ascribed to all organisms forming societies, however low they may happen to be in the scale of evolutionary organization. The lower stages of life, for example, are reported by these authors to possess 'an awareness which is extremely dim and possibly fleeting in its extent'. The old problem of ascribing different mental states to animals makes itself felt particularly strongly here, and although we may envisage lower organisms possessing some kind of awareness of the kind which we possess, this is an exercise on our part. We are attributing our own modes of seeing the world to the rest of the animal kingdom. Human characteristics are ascribed to animals from a man-centred point of view, and this often precludes adequate consideration, not only of the similarities, but also of the differences between the society of a particular group of animals and that of man.

Is there any reason to suppose that experience exists in only one type, i.e. that possessed by human beings? May there not be other forms perhaps possessed by lower organisms but not by man? It is obvious that the mentalist account is unacceptable, and that these questions must remain a matter of speculation. What is certain is that grave dangers are associated with any attempt to attribute anything like our own experiences, simply because they are our own experiences, to other organisms. There seems no point in discussing whether animal groups do or do not have mental states. Since we cannot ascertain either the existence or the nature of these states, they cannot be used as a guideline to the forces acting to cause animals to form societies or to interact with one another in any other way.

If it is argued that there is no point in discussing the mental states of animals, what can be said about the social behaviour of animals? Watson (1924) concerned himself with the problem of interpreting animal behaviour. He was forced to abandon the terminology of the mind, because it represented a different universe of discourse from the behaviour he observed. He concluded that, whilst it was impossible to say anything meaningful about the mind of a laboratory animal at this level, there was a great deal which could be said about the animal's behaviour. Watson argued that mental states are essentially private and unknowable, except to the human being who experiences them, whereas behaviour could be observed, commented upon, and theories about it could be devised and tested. If this argument is true for lower organisms, might it not also apply to man? The same problem arises in ascribing mental states to the pre-verbal human infant. We imagine the mental states of infants to be somewhat similar to our own, but we can have no knowledge that this is in fact the case. The argument which Watson put forward was that we should stop the practice of anthropomorphizing, not only with regard to lower animals, but also with regard to human infants. Arising from the study of animal behaviour, behaviourism became a full-blown discipline which attempted to apply its theoretical standpoint not only to infant but also to adult human bevaviour.

Broadbent (1965) has presented a modern account of the behaviourist approach. The study of animal behaviour in at least this respect has led to a revolution, not with regard to the specific findings about animals, but in what we conceive to be the proper methods to use in the study of mankind.

Yet another early approach rooted in the biological disciplines was that of the life and society school. This view supposes that the social behaviour of animals and man is in some way dependent on the inherent quality of life (Langer, 1958). The axiom of the life and society school appears to be that wherever there is life there is society. Life is self-perpetuating, it stems, as far as we know, from other life. Society is created by the very mechanism of heredity. Behaviour reinterpreted from this viewpoint is equated with existence. Behaviour, however, has never been the exclusive province of those disciplines which profess to be most concerned with it – ethology, psychology, etc. The engineer will talk about the behaviour of a structure, the chemist about the behaviour of a substance, the astronomer about the behaviour of a star. Behaviour means not only a principle of action with regard to an observer, but also a condition of response to the surrounding environment. There are many dangers in equating the society of man or lower animals with life. The spectre of vitalism hovers over the argument, and although the presence of

reproduction implies aggregation, at least shortly before
and after reproduction, with the advent of self-perpetuating
computer organizations the old distinctions based upon
the power of generation of the species breaks down.

The fact of life may no longer be sufficient criterion to
distinguish whether society exists or not. Matter itself
could be said to form its own society, if by this is meant the
capacity to cohere in organized groups.

By far the best documented approach arising out of the
biological disciplines to the problem of generalization between animals and man is that of the evolutionary viewpoint (Lorenz, 1963; Ardrey, 1967; Ashley Montagu,
1968). The argument can be simply stated. Man is a product of evolution. As such he shares a common heritage
with other animals. His progenitors are also those of his
closest relatives, the apes. He shares the core of his nature
with his immediate ancestors. The mechanisms of heredity
transmit to him his basic animal pattern. The similarity is
perhaps most striking in the study of developmental ana-
tomy, where man can be seen to share with his relatives
the basic vertebrate body plan. There are many similarities also in the design and function of his body, which bears
a striking resemblance to closely related vertebrate species.
If it is true that man represents physically an obvious extension of the evolutionary process, might it also be true of the
behaviour of man that it, too, is formed from an extension
of man's immediate ancestory, that it, too, is on a continuum with that of animals lower in the phyletic scale?
It is clear that we do share a bond in common with our
ancestors at, for example, the reflex level, whereby these
built-in patterns of response occur in a very similar fashion
amongst the closer relatives of man as well as in man himself. Whilst it cannot be denied that there are obvious
similarities, there are also many differences. The question

has to be asked, 'How similar must the behaviour of man be?', in order that this may be explained by 'his primitive ancestry'. Does his animal past still remain dominant within him, or has the degree to which his capacity for learning has developed, made him into a new kind of organism?

These arguments have been the cause of controversy since the time of Darwin. The case can only be regarded at the moment as non-proven, because the argument rests upon the degree to which it is supposed that man has moved along the phyletic scale. The supporters of both sides of the debate are perfectly happy to accept the concept that evolutionary theory applies to man; disagreement arises, not as to whether this has occurred or not, but as to what form it has taken. We can only regard these attempts to investigate behaviour by including the behaviour of man within an evolutionary framework, as provisional, and open to the possible objection that the form of analysis may not be appropriate to an understanding of human behaviour.

The problems of generalization

There are many difficult problems in making statements based upon observations of animal behaviour, which are intended to relate to human behaviour. These are of several types.

First, it is essential to check the accuracy of information. It may be that particular descriptions of animal behaviour have been over dramatized to secure a specific effect. It may be that recorded instances occur far less frequently than the literature would lead one to expect, because of this tendency to highlight observations. It may also be the case that accounts of animal behaviour bear little relation to the manner in which the animals actually behave. This is true of a number of early naturalistic accounts, where the boundaries between folklore and natural observation are obscure. It is sometimes assumed that the life experience of the individual person can provide the key to the nature of man. In generalizing from animals to man, it is regarded as sufficient under some circumstances to relate scientifically based animal observations, on the one hand, to the investigator's personal experience, on the other. It need not be pointed out, however, that one person's experience may be quite different from that of another. The personal experience of an author may be determined by many different factors, and individual experience is no substitute for scientific observation. Just as repeatable observation, accurate measurement and logical deduction are essential to the study of animal behaviour, so they are equally essential to the study of human behaviour, if generalization from one to the other is to be meaningful. The facts about human behaviour must also be accurately determined.

It is also possible to believe that generalizations are based on a strict logic when in fact this may not be the case. An important argument with regard to the process of generalization arises through the discipline of cybernetics, and it is argued that mathematical models or logic systems can be used to represent social behaviour or an organized social system. When social behaviour is represented by a mathematical model of this kind, when it has been expressed in terms of a logical plan, one system may appear to have a great deal in common with another, more in fact than may have been originally apparent. It is argued that, if simple models can be formulated to represent the social behaviour of animals, then they may also be used to reinterpret the nature of the behaviour of man. The argument may also be used to apply to artificial organisms. This view has a lot to recommend it, in that it concerns itself with the building of models of organic systems, and it also attempts to generalize on the ground that a common logic be used to explain different systems. These are both hallowed principles of scientific methodology. Nonetheless, the systems-theory approach suffers from serious disadvantages. In the first place, if it is to be determined that a particular model derived from the study of animal behaviour is to be applied to man, then it is necessary, at least in most major respects, to understand the behaviour of man before the model can be successfully applied, and if the model is to be anything more than a working hypothesis.

Secondly, the adequacy or good fit of a particular model depends almost entirely on the aspects of behaviour which are chosen to relate to it. The difficulty in applying to human behaviour a model which has been derived from the study of the behaviour of animals, lies in choosing which aspects of human behaviour are to be used as the raw data of the application. If there is no logical equivalence between the behaviour of animals from which the model was originally derived and the human behaviour on which the model is finally tested, then the idea that there is some inherent logic in this approach cannot be sustained. The argument once more is simply that of analogy.

Thirdly, if as it seems it is not possible on the basis of input/output data, so beloved of the system-theory model builders, to predict what the nature of the intervening arrangement is likely to be, then this introduces a principle of indeterminacy which is again the very thing that the model builders are seeking to avoid.

Perhaps the greatest problem of all, and perhaps the one which concerns us most directly is that of over-generalization. Although it is by no means a necessary function of the work which is done, the investigator of animal behaviour may in many cases wish to extend his observation and theories from animals to man as part of the biological continuum. This is a worthwhile function, because the investigator may have derived an individual perspective which does genuinely allow him to view human behaviour in a new way. The danger of over-generalization occurs, not when an investigator of animal behaviour arrives at some new viewpoint, then proposes this as a suggestion, which might be applicable to man, perhaps surveys the literature and then investigates the behaviour of man on his own account; the danger arises when the investigator states categorically that his investigations apply to human behaviour, having failed previously to become acquainted with the existing literature on human behaviour, or having failed to carry out independent investigations on human beings to establish if the statement made about their behaviour applies to them or not. A number of distinguished biologists, as the result of their extensive research with animals, are at present making statements based on their research which they believe to apply to man. In some cases they have ignored the literature stating what has already been discovered about the behaviour of man, and they have not always investigated the behaviour of man in order to establish the application of their views.

What seems to be essential if generalizations are to be made is that adequate parallel investigations should be made of human behaviour. It is not useful to apply elaborate scientific procedures to the observation of animal behaviour if the ideas derived from this are applied to man purely on the basis of the *ad hoc* personal experience of the investigator. The ideas need elaborate authentic verification. The ideas must stand the test and they must rely on objective evidence. In many cases investigations using human subjects are possible, but in some other cases animal research is undertaken because the investigation cannot be conducted at the human level. Even if this is accepted, there is still no excuse for not sifting the evidence, although it may go only part of the way to answering the specific question which has been asked about human behaviour.

Finally, some of the difficulties which are generated by an over-zealous application of the findings from studies of animal behaviour to man arise because of a failure fully to recognize the nature of species differences. Animals of one particular species may behave very differently from animals of other closely related species. In many ways homo sapiens behaves in a very different way from the most closely related primates. Any attempt to express views which fail to recognize these species differences may be in danger of appearing absurd. It is important, therefore, that full recognition be given to the difference which exists between man and his closest relatives. The implication of this is that specific patterns of behaviour do not necessarily translate from one animal to another. Behaviour patterns are used for very different purposes, and similar patterns may stem from very different origins. The attempt to compare directly the behaviour patterns of man and lower animals may appear facile and misleading, but there are more important things to compare than this. These are the broad general principles of behaviour: the study of causes which lie beyond specific behaviour patterns themselves; the study of determinants, the study of endogenous and exogenous effects. Most important of all are the ideas about animal behaviour, the theories of animal function. It is these which ultimately may be capable of changing our perspective if properly applied to human life and conduct.

Applications to the behaviour of man

One of the most difficult areas in which it is possible to work concerns the nature of the application of research findings gathered from animals to man. We can now consider how the research reported in this book could fruitfully be applied to the study of man. It is possible to suggest some areas of investigations in which it is hoped that the arguments based upon animal research could be taken up and extended into the human sphere.

The study of early experience has received wide currency in the animal-behaviour literature. The study of sensory deprivation has been reported in primates to have dramatic effects. This work has been taken up by a number of investigators of human infant behaviour (Bowlby, 1958), and independent investigation undertaken at the human level.

The concept of imprinting has been similarly employed. Research findings on infra-human animals have been compared with, for example, the smiling response in human infants (Spitz and Wolf, 1946; Salzen, 1963). The search for analogous patterns of behaviour in the human infant to that of neonate chicks or ducklings seems, however, to have dogged this application to human behaviour and to have rendered the analogy far less acceptable than it must otherwise have been.

Some of the most interesting work relates to the development of a particular preference for objects to which the infant has been exposed during the early period of its life. This work really can be classified as exposure learning, although it becomes inextricably bound up with the concept of imprinting (Sluckin, 1964). The development of sexual preference in animals as the result of early experience is one of the most important of the findings. Preference, however, seems to develop for many different objects to which the animal has been exposed: food preference, home run preference, object preference, companion preference. All these things point to the importance of early experience in laying down the guide lines of subsequent choice. This is an area of research which needs development at the human level. Some attempts have been made to examine object and social choice as the result of different kinds of early experience, but by and large these attempts have been largely anecdotal and unsatisfactory in the methods used.

Embryological studies are of great importance in our understanding of the development of human behaviour. It is apparent that much of the primitive development of behaviour has already taken place by the time the mother gives birth to the infant. The study of the development of the behaviour in the human embryo is a difficult and complicated research area. It is to be hoped that the advent of sophisticated electronic apparatus will eventually bring many more investigations within the realm of possibility, but meanwhile we are forced to rely almost entirely on the relatively small number of investigations of behaviour on various non-human animal groups.

The study of the development of behaviour in animals is of course of great importance to the study of the development of behaviour in man. The development of learning, of perception, of problem solving, of motor capacities are all interrelated areas which have their companion studies in human behaviour (Sluckin, 1964).

The study of family processes in animals is perhaps one of the major areas where ideas about the human family can be clarified by an examination of the nature of animal behaviour. The processes of maternal behaviour as studied in laboratory animals could throw light on the nature of maternal behaviour in human beings. The influence of hormone systems and the endogenous changes which come about in maternal response, as well as the effect of the stimuli emitted by the pup, could all be regarded as a primitive maternal system which may be analogous to the human one. The fact remains that, although cultural differences of a wide variety exist between different ethnic groups, mothers do protect and look after their children. It is not difficult to envisage experiments at the human level designed to investigate physiological changes associated with mothering, endogenous changes and those stimuli which call forth maternal responses in the human female.

The analysis of sexual behaviour in animals appears to be particularly apposite to that shown by human beings. The investigations of sexual behaviour in human beings is not easy. However, many of the experiments designed to investigate sexual behaviour in animals need to be carried out at some time on human beings in order for the parameters of sexual response to be determined. Is it possible to identify the appropriate stimuli to sexual performance in human beings? What are the nature of the endogenous changes which are associated with sexual response? What are the effects of the sex hormones, not upon the physical structures, but upon the nature of sexual behaviour? What are the limits of sexual functions? These are all questions which now need to be investigated at the human level.

Aggression in man is one of those areas which needs particularly close attention. It is necessary to study under strictly controlled experimental conditions the type of aggression shown by children. It is necessary to examine its use as an instrumental act, its appearance under conditions of overcrowding and frustration, its manifestation to certain stimuli such as those associated with pain. Only then can evidence be obtained to suggest whether aggression is a reactive process or one lingering on from man's evolutionary past. Hutt and McGrew (1968) have begun appropriate studies of this kind, but it is obvious that much greater attention needs to be devoted to this area.

The study of social motivation in animals points to means by which this important topic might also be studied in human beings. In the first place, methods of this kind are being used to study the social behaviour and motivation of preverbal infants; secondly, more sophisticated investigations using somewhat similar methods could be employed upon children and adults. The possibility arises from the study of avoidance learning in animals that fear communicates itself from one animal to another, and this is sufficient in the absence of any other kind of reward or punishment to produce highly effective learning. It is obvious from this study that the activities of the group engender various forms of learning in group members without obvious forms of reinforcement being present. This infectious learning is something which needs to be far more widely studied at the human level, and here the experiments on animals would seem dramatically to point the way to this type of investigation.

Extensions of operant behaviour

In this book we have been concerned to an extent with the experimental analysis of social behaviour in animals. It will be remembered that by this is meant the use of operant methods to study social behaviour. The use of operant methods represents an extension of the desire to be objective wherever possible, and to use precise measures. This movement was initially devoted almost entirely to the study of learning processes in animals. Out of it has grown, however, several major approaches to the study of human behaviour. Foremost of these is the contribution to the study of learning in human beings. This is particularly important with regard to understanding the nature of learning in children. In the investigation of operant studies of learning in animals lie the origins of the automated instruction and the educational technology movement which introduced sweeping changes into educational methods. These influences have not been without effect in adult education and, in particular, with respect to the rapid acquisition of new skills in industrial training. Conditioning studies are contributing widely to various forms of behaviour therapy in the clinical treatment of a range of behavioural disturbances. It is not the object of this chapter to elaborate upon these achievements, although their origins lie so clearly in the study of animal behaviour. The aim is to illustrate another development stemming from the same operant technique, but one which relates specifically to the investigation of social behaviour.

Skinner (1962) began a new kind of discipline when he first used operant response to study social behaviour in animals. He used two pigeons and required them to coordinate their response in order to receive food reward. Since that time, social behaviour has been investigated by these techniques in a wide variety of species. These investigations have been fully described in previous chapters. Here again, in the application of research findings obtained on animals to man, we wish to concentrate not on the specific findings but on the general principles evolving from them. We know that man behaves in a socially integrated fashion, we know that social rewards are important to him, we know that he is capable of a great deal of cooperation and altruism, as well as of competition with his fellows. What we do not really understand as yet is how these particular patterns of behaviour come about, and how it is that they occur under some circumstances but not others. The suggestion has been made that society is its own conditioning device. Man is peculiarly sensitive to conditioning, and social conditioning is a powerful tool. Each society carries out its own conditioning programme. The argument that we are each conditioned by the social influences around us is not a new one. Recently, however, this idea is being expressed in a new way in terms of an operant-box sociology movement. Animals placed in operant apparatus are under particular constraints. Their movement is somewhat restricted by the apparatus, and the nature of the reinforcement they receive is determined by the contingencies of the apparatus. The whole environment is a training box. The environment itself acts to promote learning, both by the nature of the constraints it places upon the organism, and by the nature of the contingencies of the reinforcement it provides. Just as a gymnasium might be described as a sweat box, so a piece of operant apparatus can be described as a learning box. The essential concept is that of the environment acting upon the organism, not only to determine what it is allowed to learn, but also the contingencies of the reinforcement by which learning is stamped in. The operant box is a piece of apparatus which allows control to be established over the behaviour of the organism.

These concepts of the operant box can be applied to the social behaviour of man. Interest is expressed in this case, not in the fine detail of animal learning, but in the general conditions which promote learning. Can these be equated for lower animals and man? The suggestion which arises is that society, with regard to the behaviour of its human members, acts as the equivalent of the operant box. In the first place, society imposes constraints upon each individual. The person is not free to behave according to his every whim. His range of behaviour is determined by the social conditions around him, by the needs and demands of other people, as well as his own needs and demands. The individual is essentially constrained by this environment.

Society does more than this. Society holds the supply lines of reinforcement to the individual, just as the contingencies of the operant box determine whether reward will be administered or withheld, or how much reward will be given. The individual receives reward from his environment. The contingencies of this are determined by the environment. It is the environment which calls the tune. Reward can be of two types, both positive and negative. Negative reward may be as influential as positive in determining the cause of behaviour. The fear of punishment acts as a powerful motivating force, as well as the fear that positive reward will be withdrawn. The social rewards provided by others, as well as the opportunity to be with others can provide a powerful incentive. The avoidance of areas of danger can be learnt and socially communicated from one animal to another, without that animal ever actually experiencing the danger. Evidence of this kind points to the importance of social reward, social reinforcement and social learning as an influence on the behaviour of organisms. There is every reason to suppose that these factors are as powerful, if not more powerful, in determining the nature of human response. Social reinforcement is regulated by society itself, and this again appears to be a major control over the supply lines of reinforcement meted out to the individual.

Finally, the operant box is a learning environment. It teaches the organism how to behave. The operant-box sociology approach supposes that each social environment in which a human being finds itself is also a teaching environment. The contingencies set by the environment teach the individual some things rather than others. The environment determines what each individual learns. The family, the school, the play group, the adolescent gang, the office, the prison, the mental hospital, each provide an operant box to the individual who learns the behaviour which that institution sets down. The schedule prescribed by the management of an institution may be totally different from that to which the individual is exposed. The delinquent, for example, may undergo a series of formal lessons prescribed by the educational system of the establishment, yet these may play a little part in terms of the reinforcement and training of the environment as an operant box. The rewards provided by the acclaim of classmates may form a much more powerful teaching device than that of the formal instruction from above. The institution itself is providing in this case a powerful operant teaching system which is quite different in scope and kind from the one the management envisages it as being. Whatever the nature of delinquency it is not our function to discuss its causes here; it is used to illustrate the extension of ideas gathered from studies of operant conditioning in animals to the question of the social organization of man, and is illustrative of the way in which concepts and not specific results necessarily can be used to further our understanding of the nature of social organization.

Summary

The broad principles and the criteria by which it is possible to generalize from animals to man are discussed. Historically, this problem has been approached through mentalistic and anthropomorphic viewpoints. There are many problems associated with these, particularly the old one of ascribing mental states to animals and particularly those low in the phyletic scale. The life and society viewpoint also has dangers. Arguments are put forward to show that the fact of life is no longer sufficient criterion to distinguish whether society exists or not. The behaviourists and the evolutionary viewpoints are discussed. The behaviourists position is supported because of the value of its methods, but the evolutionary viewpoint is regarded as non-proven.

The problems of generalization relate to the accuracy of information at the human level. If the comparative argument is to be followed, the accuracy of information obtained at the human level must match that obtained from animals. The systems-theory approach also depends on an accurate knowledge of human behaviour, as well as appropriate application of the findings from the study of animal behaviour to man. Over-generalization is one of the greatest problems. It is important that full recognition be given to the difference which exists between man and his closest relatives. It is not often possible to translate specific patterns of behaviour from one animal to another. Behaviour may be used for very different purposes. The attempt to compare directly the behaviour patterns of man and lower animals may appear facile, and indeed there are more important things to compare – for example, the broad general principles and the theories of animal function.

Many aspects of the work concerned with the study of the social behaviour of animals could seek useful application to the study of man. These aspects are in the realm of early experience, development of object preference, embryological studies, behavioural development, family and group processes, parental and sexual behaviour, and so on.

The general principles of the study of operant behaviour in social situations also has application to the study of human behaviour. Social conditioning is an important process and recently the idea has been expressed in a new way through the operant-box sociology movement, whereby society provides constraints and reinforcements which determine and shape the behaviour of those members who contribute to them.

After reviewing the various areas of study of the social behaviour of animals, we can discuss the problems in terms of different kinds of factor. First, we have to discuss the physiological processes which relate to behaviour, the structure of the nervous system, endocrine function and the balance achieved by different hormone systems, and so on. These factors are concerned with the somatic approach to social behaviour and they place it firmly in the biological context.

The second group of factors are the ones which concern the interrelations between one animal and another, the nature of communication systems and the function of social stimuli, in addition to other facets of group behaviour. These factors can be described as socio-behavioural.

The question we now have to ask concerns the interrelationships of these two kinds of factor. How do physiological and socio-behavioural factors work together? How do external events come to relate to the processes at work inside the organism?

One suggestion is that the internal events represent a potential which can be called upon as the demands of the environment dictate. This classes the physiological processes organizing behaviour as a reserve, and the organism as a reactive system moving to the demands of the external world. This is analogous to the instrumentalists who meet together in an orchestra. Each instrumentalist takes part only at those times when his performance is called upon by the score, and in this case the score represents stimuli from the external world. The capacity to play still exists, although the score may not demand it, and so the potential remains. This analogy cannot be carried very far, however, because although some systems appear to do little more than to react to the events of the environment, some other systems are more active agents than that. Not only do they provide a potential for performance, but they also call the tune.

Endogenous factors, for instance, relate closely to social drive. They can be regarded as the mainspring of the individual animal's performance, and in experiments on the effects of early isolation on subsequent sexual behaviour it has been demonstrated (Beech, 1968) that some animals may show disturbed sexual behaviour, not through any change in sexual drive, but through an inability to orientate properly with regard to the sexual partner. Indeed, mounting may occur as frequently as in normal animals, but the success of mating is considerably reduced. Sexual drive in this case appears to be relatively unaffected by the effects of early isolation, but the control of behaviour by the events of the environment does. This highlights one relationship between internal and external processes, in which environmental stimuli may be important in directing and orienting the behaviour, which may be triggered from internal and hidden forces. These form the mainspring of the behaviour, but the socio-behavioural factors are responsible for guiding and channelling the response.

It is sometimes argued that the concept of drive is an unnecessary one and that an active organism will be led to behave in the way that it does by the nature of the control exercised over it by environmental factors. This view, however, whilst it may form a starting point for a discussion of motivation and may be invoked as an explanation of neonate behaviour, does not fully accord with the evidence for the specific effects in adult life of different drive states, as, for example, in the association of different hormones with particular behavioural states.

Physiological events may also perhaps exert their effects through the functions of the perceptual systems. They may lead the organism to become much more sensitive or they may introduce selectiveness, causing intense response to some stimuli, whereas others are totally ignored. The evidence, for example, of monkey viewing of infants by their mothers suggests that drive states, maternal drives in this case, influence the perceptive process at least in terms of those objects which are preferred.

The model of drive which this implies is one whereby one particular drive under the influence of endogenous processes takes over the perceptual system, and this leads to a selective response. Some parts of the environment will be chosen as providing signals for response, whereas others will not. When the drive state changes, different influences could come to bear on the perceptual systems. Different filters will be switched in, and what is selected for response may now not be the same as that selected previously. Response to an object ignored previously may now occur, because under the fresh influence of physiological events the animal now becomes sensitive to it. Male monkeys, for example, may associate with the female in some species when she is on heat, but not at other times.

These processes may act, not only upon the perceptual systems to change their sensitivity and to introduce a principle of selection, but also on behaviour itself. It is well-known that hormone concentrations influence the type and intensity of the behaviour of the animal to which they are administered, and it may well be that endogenous factors are responsible for the selection of some behaviour patterns rather than others. This has more far-reaching effects than may at first seem, because in acting directly upon behaviour to select out some patterns rather than others, endogenous factors could work through behaviour to predetermine the nature of social interaction. Physiological events may lead the animal to display some patterns rather than others. These are not without their effect upon other animals, and these patterns in their turn influence the behaviour of other animals. The nature of the behaviour of the original and the other partners determines the nature of the social interaction. Communication is important in this respect, and one of the major influences may be exercised through the types of behaviour which are used for this purpose.

This discussion of physiological events in their relation to the social behaviour of animals sets out to show that each animal is both a reactive agent, in the sense that it responds to the behaviour of other animals, and also a determining agent, whereby through the nature of its own drives and behaviour it leads and influences the behaviour of others.

It is easy to create an artificial division between those who emphasize the reactive nature of social response, on the one hand, and those who emphasize the importance of physiological drive on the other. This happens largely because it is supposed that the organism has to be one thing or the other.

All the evidence suggests that it fulfils both of these functions. It is quite possible that different mechanisms are involved, but nonetheless the internal and external events are chained into the responses of the one organism. They influence one another, and mutually interact, and this means that the system is so formulated that the internal factors establish priority at one time, whereas the external ones lead at another. The organism is both a reactive and a determining agent, and the way its behaviour is to be interpreted depends upon the state which it has assumed at a particular time.

Whilst discussion of endogenous processes is essential to a full consideration of the nature of the social behaviour of animals, the argument cannot proceed very far if these factors are considered in isolation. No complete explanation can be provided by discussion of these factors alone. An examination of neural structures or endocrine balance tells us very little about social behaviour, if it is not fully related to the nature of the behaviour which it is supposed to explain. The question of how endogenous and exogenous factors relate to one another now has to be considered in socio-behavioural terms. Before doing this, however, attention must be drawn to the influence which external events may have upon physiological events. It is perhaps a mistake to think that physiological events only influence the response to the events of the outside world, and that this is a one-way affair. Socio-behavioural events also exert a profound influence upon the nature of the internal physiological processes. The hormone systems not only exert an effect upon behaviour but they are themselves influenced by the events of the external environment. This has been elegantly demonstrated by Lehrman and Erickson (1962), who showed that the visual stimulation provided by an adult male pigeon to a female pigeon in an adjacent cage was responsible for a considerable increase in her oviduct weights. The visual stimulation provided by a castrated male also had some influence in increasing the oviduct weights of the female, but not to the same extent as the normal animal. This work illustrates the effect which environmental stimuli may have upon internal morphological characteristics and endogenous function. Another experiment which illustrates the fact that internal organ weights may be altered by the nature of environmental stimuli, is that of Krech, Rosenzweig and Bennett (1962), who showed that rats reared in an enriched environment show an increase in brain weight over those reared in an impoverished environment. This illustrates something of the reactive and adaptive nature of the endogenous processes themselves.

One of the important ways in which environmental events may be linked to important internal processes is through conditioning. An environmental stimulus becomes associated with an established bodily response in classical conditioning. The dog now produces saliva to the sound of the buzzer, instead of in response to the presence of food. Recently, the importance of conditioning of the autonomic nervous system has been emphasized. It was formerly supposed (Mowrer, 1960) that classical conditioning related to the autonomic nervous system, and operant conditioning to the central nervous system. Trowill (1967) and Miller and Carmona (1967) have shown, however, that operant conditioning of autonomic processes is perfectly possible. They conditioned autonomic systems by reinforcing any change in heart rate in a particular direction, either slower or quicker, with electrical stimulation to the hypothalamic reward areas. Experiments of this sort illustrate clearly that conditioning of the autonomic nervous system can be established. It has long been suggested that autonomic conditioning may be related to the social effects of fear and anxiety because, of course, the bodily changes associated with emotional behaviour of this kind are so closely under autonomic control, and could so easily be called out as learned responses to social stimuli with which they have become associated.

The significance of this with regard to social behaviour is that not only may behaviour patterns themselves become conditioned, but so also may the endogenous processes to the social stimulus provided by other animals. This applies particularly to the central nervous system, but also to the autonomic nervous system, and the work of Lehrman raises the interesting possibility that it may apply to hormone and other systems as well. One animal may act as a stimulus, not only for the conditioned behaviour of another, but also for the conditioned response of many different internal events of the other.

The investigations described in this book point to the fact that operant conditioning is far more than a simple technique for studying learning. Indeed, we are so accustomed to think of it as a means of studying learning that it is often forgotten that it is also a means of studying biological factors in behaviour and, more than that, a method which allows us to say something important about social behaviour. This can be substantiated as the result of many different investigations, and these have been fully described previously. A few examples are quoted in order to draw the work together and in order to show how problems in different areas of social inquiry can be tackled by the use of these methods.

Skinner (1962) in his experiments on cooperation amongst pigeons showed that animals can be made to demonstrate a high degree of sensitivity to one another. They are by no means the dumb brutes guided by only irrational instinct that some of the philosophers would have us believe. This sensitivity has been illustrated in many other investigations. In experimental investigations of the maternal behaviour of primates, for example, when female monkeys are deprived of their infants they become extremely agitated, they roam around the cage and give every appearance of searching for their lost young.

Cross and Harlow (1963) showed that the mother, when separated from her infant, will work to open a window to look through at her infant. The degree to which she is prepared to do so provides an index of her maternal response. She opens the window more frequently to view her second- or third-born infants than her first-born, and this kind of experiment illustrates the use of operant methods to study the waxing and waning of maternal response. These methods allow us to know what aspects of social life are rewarding to the animals, as well as to study the nature of endogenous changes in maternal response.

Less familiar are the studies of the conditioning of socio-sexual behaviour in animals; Schwartz (1956), for example, trained male rats to press a lever to secure a sexually receptive female and in this way was able to study the effects of deprivation and satiation on male sexual response. Butterfield (1969) also has used methods of this kind to study sexual preference and the development of the pair bond amongst finches. Methods of this kind are being used to study important aspects of sexual behaviour, such as the development of the preference for one animal rather than another, as well as difficult areas like the development of sexual anomalies and sexual pathology.

We are just beginning to see the development of operant applications to the study of aggressive behaviour. In view of the importance of aggression in man it is a very significant problem, not only as to how aggression develops, but also as to how it may be suppressed. Patterns of aggressive behaviour can be easily conditioned by operant means. If, for example, a pigeon receives food reward every time it pecks another bird, it soon becomes a highly aggressive animal, not only in this but in other situations (Willis *et al.*, 1966). These experiments and others concerned with the suppression of aggression seem to have considerable potential in providing a model for aggressive behaviour which could perhaps be subsequently applied at the human level.

One of the striking features of much animal behaviour is the degree to which animals are prepared to cooperate with one another. Church (1959) showed that rats are highly sensitive to the pain experienced by other rats. A rat pressing a lever to obtain pellets of food would reduce the rate of lever pressing, when another rat suffered pain because it had been given electric shock. Rice (1965) found that, if one rat had been immersed in water, another animal very quickly came to its aid. Hansen and Mason (1965) showed that a monkey will stop pressing a lever to obtain food reward if that lever also activates a circuit which is used to provide another monkey with electric shock. These are examples of the use of operant methods to study the important social interactions between one animal and another. Animals are not able to tell us in our language what they think or feel. Operant methods seem to be one way of by-passing the language barrier. These methods can be used not only to control particular patterns of behaviour but also to demonstrate the social preference of animals and to say something of the nature of morality in social relationships.

The significance of the operant movement with regard to studies of social behaviour is now only just beginning to be fully appreciated. It has a major contribution to make to the study of the social behaviour of animals. Just as we have observed the growth of the ethological movement into the full-blown discipline of ethology, so it would seem we are witnessing the development of the operant approach into an integrated operant behavioural science. Operant conditioning is far more than a technique for studying learning; it has a major contribution to make to the development of behavioural science and to those areas which are concerned with the problems of social behaviour and development.

References

Ader, R., Beels, C. C., and Tatum, R. (1960). Social factors affecting emotionality and resistance to disease in animals. II, Susceptibility to gastric ulceration as a function of interruptions in social interactions and the time at which they occur. *J. comp. physiol. Psychol.*, 53, 455–8

Allee, W. C. (1931). *Animal Aggregations*, University of Chicago Press

Allee, W. C., Collias, N. E., and Lutherman, C. Z. (1939). Modification of the social order in flocks of hens by the injection of testosterone propionate, *Physiol. Zool.*, 14, 412–40

Allen, M. D. (1957). Observations on honeybees examining and licking their queen. *Brit. J. Anim. Behav.*, 5, 81–4

Altman, I., and Haythorn, W. W. (1967). The ecology of isolated groups, *Behavioural Science*, 12(3), 169–82

Amsel, A. (1962). Frustrative non-reward in partial reinforcement and discrimination learning. Some recent history and a theoretical extension, *Psychol. Rev.*, 69, 306–28

Anand, B. K., and Brobeck, J. R. (1951). Hypothalamic control of food intake in rats and cats, *Yale J. biol. Med.*, 24, 123–40

Angermeier, W. F. Schaul, L. T., and James, W. T. (1959). Social conditioning in rats, *J. comp. physiol. Psychol.*, 52, 370–2

Angermeier, W. F. (1960). Some basic aspects of social reinforcements in albino rats, *J. comp. physiol. Psychol.*, 53, 364–7

Angermeier, W. F. (1962). The effect of a novel and novel noxious stimulus upon the social operant behaviour in the rat, *J. genet. Psychol.*, 100, 151–4

Angermeier, W. F., Philhour, P., and Higgins, J. (1965). Early experience and social groupings in fear extinction of rats, *Psychol. Rep.*, 16(3, pt. 2), 1005–10

Angermeier, W. F., Phelps, J. B. Reynolds, H. H., and Davies, R. (1967). Performance and biochemical responses related to social changes versus chemotherapy in nonhuman primates (rhesus monkeys), USAF ARL, *Tech. Rep.*, 67, 23

Angermeier, W. F., and Phelps, J. B. (1967). Dominance in monkeys: behaviour and biochemistry. USAF ARL, *Tech. Rep.*, 67, 24, v. 13

Angermeier, W. F., Phelps, J. B., Murray, S., and Reynolds, H. H. (1967). Dominance in monkeys. Early rearing and home environment, *Psychonomic Science*, 9(7B), 433–4

Angermeier, W. F., Phelps, J. B. Reynolds, H. H., and Davies, R. (1968). Dominance in monkeys: effects of social change on performance and biochemistry, *Psychonomic Science*, 11(5), 183–4

Archer, J. (1970). Effects of population density on behaviour in rodents, *Social Behaviour in Animals and Man*, ed. J. H. Crook, Academic Press, New York

Ardrey, R. (1967). *The Territorial Imperative*: a personal enquiry into the animal origins of property and nations, London

Ashley Montagu, M. F. (1968). *Man and Aggression*, Oxford University Press

Azrin, N. H., Hake, D. F., and Hutchinson, R. R. (1963). Elicitation of aggression by a physical blow, *J. exp. Anal. Behav.*, 8(1), 55–7

Azrin, N. H., Hutchinson, R. R., and Hake, D. F. (1963). Pain-induced fighting in the squirrel monkey, *J. exp. Anal. Behav.*, 6, 620–1

Azrin, N. H., Hutchinson, R. R., and Hake, D. F. (1966). Extinction induced aggression, *J. exp. Anal. Behav.*, 9(3), 191–204

Baenninger, L. P. (1966). The reliability of dominance orders in rats, *Animal Behav.*, 14(2–3), 367–71

Baenninger, R. (1967). Contrasting effects of fear and pain on mouse killing by rats, *J. comp. physiol. Psychol.*, 63, 298–303

Bambridge, R. (1962). Early experience and sexual behaviour in the domestic chicken, *Science*, 136, 259–60

Bard, P. (1928). A diencephalic mechanism for the expression of rage with special reference to the sympathetic nervous system, *Am. J. Physiol.*, 84, 490–515

Bard, P. (1934). On emotional expression after decortication with some remarks on certain theoretical views, Parts i and ii, *Psychol. Rev.*, 41, 309–29, 429–49

Barnett, S. A. (1963). *A Study in Behaviour*, Methuen, London

Baron, A., and Littman, R. A. (1961). Studies of individual and paired interactional problem solving behaviour of rats. II, Solitary and social controls. *Genet. Psychol. Monogr.*, 64, 129–209

Bateson, P. P. G., and Reese, E. P. (1968). Reinforcing properties of conspicuous objects before imprinting has occurred, *Psychonomic Science*, 10(11), 379–80

Bayer, E. (1929). Beiträge zur Zweikompanenten Theorie des Hungers, *Z. Psychol.*, 112, 1–54

Beach, F. A. (1940). Effects of cortical lesions upon copulatory behaviour, *J. comp. Psychol.*, 29, 193–244

Beach, F. A. (1941). Copulatory behaviour of male rats raised in isolation and subjected to partial decortication prior to the acquisition of sexual experience, *J. comp. Psychol.*, 31, 457–70

Beach, F. A. (1942). Comparison of copulatory behaviour of male rats raised in isolation, cohabitation and segregation *J. genet. Psychol.*, 60, 121–36

Beach, F. A. (1942). Analysis of the stimuli adequate to elicit mating behaviour in the sexually inexperienced male rat, *J. comp. Psychol.*, 33, 163–207

Beach, F. A. (1945). Hormonal induction of mating responses in a rat with congenital absence of gonadal tissue, *Anat. Rec.*, 92, 289–92

Beach, F. A., and Gilmore, R. (1949). Response of male dogs to urine from females in heat, *J. Mammal.*, 30, 391–2

Beach, F. A., and Holz-Tucker, A. M. (1949). Effects of different concentrations of androgen upon sexual behaviour in castrated male rats, *J. comp. physiol. Psychol.*, 42, 433–53

Beach, F. A. (1952). Instinctive behaviour: reproductive activities, in *Handbook of Experimental Psychology*, ed. S. S. Stevens, Wiley, New York

Beach, F. A., and Jaynes, J. (1956). Studies of maternal retrieving in rats. II, Effects of practice and previous parturitions, *Amer. Nat.*, 90, 103–9

Beach, F. A. (1958). Normal sexual behaviour in male rats isolated at fourteen days of age, *J. comp. physiol. Psychol.*, 51, 37–8

Beach, F. A., and Rabadeau, R. G. (1959). Sexual exhaustion and recovery in the male hamster, *J. comp. physiol. Psychol.*, 52, 56–61

Beach, F. A., and Whalen, R. E. (1959). Effects of intromission without ejaculation upon sexual behaviour in male rats, *J. comp. physiol. Psychol.*, 52, 476–81

Beach, F. A. (1968). Coital behaviour in dogs. III, Effects of early isolation on mating in males, *Behaviour*, 30(2–3), 218–38

Bennett, E. L., Diamond, M. C., Krech, D., and Rosenzweig, M. R. (1964). Chemical and anatomical plasticity of the brain, *Science*, 146, 610–19

Bermont, G. (1961). Response latencies of female rats during sexual intercourse, *Science*, 133, 1771–3

Bermont, G. (1964). Effects of single and multiple enforced copulatory intervals on the sexual behaviour of male rats, *J. comp. physiol. Psychol.*, 57(3), 398–403

Bermont, G. (1965). Rat sexual behaviour: photographic analysis of the intromission response, *Psychonomic Science*, 2(3), 65–6

Bermont, G., and Westbrook, W. (1966). Peripheral factors in the regulation of sexual contact by female rats, *J. comp. physiol. Psychol.*, 61(2), 244–50

Bernstein, I. S. (1964). Group social patterns as influenced by removal and later reintroduction of the dominant male rhesus, *Psychol. Rep.*, 14(1), 3–10

Bevan, W., Dawes, W. F., and Levy, G. W. (1960). The relation of castration and androgen therapy and pretest fighting experience to competitive aggression, *Anim. Behav.*, 8, 6–12

Biernoff, A., Leavy, R. W., and Littman, R. A. (1964). Dominance behaviour of paired primates in two settings, *J. Abn. and Soc. Psychol.*, 68(1), 109–13

Bingham, H. C. (1928). Sex development in apes, *Comp. Psychol. Monogr.*, 5, 1–165

Birch, H. G. (1956). Sources of order in the maternal behaviour of animals, *Amer. J. Orthopsychiat.*, 26, 279–84

Blauvelt, H. (1955). Neonate mother relationships in goat and man, *Group Processes*, Josiah Macy Jr. Foundation, New York

Boelkins, R. C. (1967). Determination of dominance hierarchies in monkeys, *Psychonomic Science*, 7(9), 317–18

Bolles, R. C., and Woods, P. J. (1964). The ontogeny of behaviour in the albino rat, *Animal Behav.*, 12, 427–41

Bolles, R. C., Rapp, H. M., and White, G. G. (1968). Failure of sexual activity to reinforce female rats, *J. comp. physiol. Psychol.*, 65(2), 311–13

Bowlby, J. (1951). *Maternal Care and Mental Health*, World Health Organization, Geneva

Bowlby, J. (1957). Symposium on the contribution of current theories to an understanding of child development, *Brit. J. med. Psychol.*, 30, 230–40

Braddock, J. C., and Braddock, Z. I. (1955). Aggressive behaviour among females of the Siamese fighting fish, *Betta splendens*, *Physiol. Zool.*, 28, 172

Braddock, J. C., and Braddock Z. I. (1958). Effects of isolation and social contact upon the development of aggressive behaviour in the Siamese fighting fish, *Betta splendens*, *Animal Behav.*, 6, 249

Brian, M. V. (1957). Caste determination in social insects, *Ann. Rev. Ent.*, 2, 107–20

Broadbent, D. E. (1956). *Perception and Communication*, Pergamon Press, London

Broadbent, D. E. (1965). *Behaviour*, Eyre & Spottiswoode, London

Brooks, C. Mc. C. (1937). The role of the cerebral cortex and of various sense organs in the excitation and execution of mating activity in the rabbit, *Amer. J. Physiol.*, 120, 544–53

Bruce, H. M. (1961). Time relationships in the pregnancy block induced in mice by strange males, *J. reprod. Fertil.*, 2, 183–242

Brvce, H. M., and Parrot, D. M. V. (1960). Role of olfactory sense in pregnancy block by strange males, *Science*, 131, 1526

Burdina, V. N., Krasuskiĭ, N., and Chebykin, D. A. (1960). K. voprosi o zavisimosti formirovaniia vyssheĭ nervnoĭ deiatel'nosti sobak otuslovii ikh vospitaniia v ontogeneze, *Zh. Vyssh. Deiatel.*, 10 427–34

Butler, C. G. (1967). Insect pheromones, *Biol. Rev.*, 42, 88–130

Butterfield, P. (1970). An analysis of the pair bond in the zebra finch, in *Social Behaviour in Animals and Man*, ed. J. H. Crook, Academic Press, New York

Calhoun, J. B. (1962). Population density and social pathology, *Scientific American*, Feb. 1962, 139–48

Cambell, B. A., and Pickleman J. R. (1961). The imprinting object as a reinforcing stimulus, *J. comp. physiol. Psychol.*, 54, 592–6

Carbaugh, B. T., Schein, M. W., and Hale, E. B. (1962). Effects of morphological variations of chicken models on sexual responses of cocks, *Animal Behav.*, 10(3–4), 235–8

Carpenter, C. R. (1940). A field study in Siam of the behaviour and social relations of the gibbon, *Comp. Psychol. Monogr.*, 16, 38–206

Carpenter, C. R. (1942). Sexual behaviour of free ranging rhesus monkeys (*Macacca mulatta*), *J. comp. Psychol.*, 33, 113–62

Carr, W. J., and Caul, W. F. (1962). The effect of castration in the rat upon the discrimination of sex odours, *Animal Behav.*, 10(1–2), 20–7

Carson, J. A. (1967). Observational learning of a lever pressing response, *Psychonomic Science*, 7(12), 197

Carter, G. S. (1951). *Animal Evolution: A Study of Recent Views of Its Causes*, Sidgwick & Jackson, London

Christian, J. J. (1959). Lack of correlation between adrenal weight and injury in grouped male albino mice, *Proc. Soc. Exp. Biol. Med.*, 101, 166–8

Christopherson, E. R., and Wagman, W. (1965). Maternal behaviour in the albino rat as a function of self-licking deprivation, *J. comp. physiol. Psychol.*, 60(1), 142–4

Church, R. M. (1959). Emotional reactions of rats to the pain of others, *J. comp. physiol. Psychol.*, 52, 132–4

Church, R. M. (1961). Effects of a competitive situation in the speed of response, *J. comp. physiol. Psychol.*, 54, 162–6

Collias, N. E. (1943). Statistical analysis of factors which make for success in initial encounters between hens, *Amer. Nat.*, 77, 519–38

Collias, N. E., and Joos, M. (1953). The spectrographic analysis of sound signals of the domestic fowl, *Behaviour*, 5, 175–87

Collias, N. E. (1956). The analysis of socialization in sheep and goats, *Ecology*, 37, 228–39

Collias, N. E., and Collias, E. C. (1956). Some mechanisms of family integration in ducks, *Auk.*, 73, 378–400

Craig, W. (1914). Male doves reared in isolation, *J. Anim. Behav.*, 4, 121–33

Crane, J. (1941). Crabs of the genus Uca from the west coast of Central America, *Zoologica. N.Y.*, 26, 145–207

Creer, T. L., Hitzing, E. W., and Schaeffer, R. W. (1966). Classical conditioning of reflexive fighting, *Psychonomic Science*, 4(3), 89–90

Crook, J. H. (1961). The basis of flock organization in birds, in *Current Problems in Animal Behaviour*, ed. Thorpe and Zangwill, Cambridge University Press

Cross, H. A., and Harlow, H. F. (1963). Observations of infant monkeys by female monkeys. *Perceptual. Mot. Skills*, 16, 11–15

Daniel, W. J. (1942). Cooperative problem solving in rats, *J. comp. Psychol.*, 34, 361–8

Daniel, W. J. (1947). How cooperatively do individual rats solve a problem in a social situation? *Amer. Psychologist*, 2, 316 (abstract)

Darby, C. L., and Riopelle, A. J. (1959). Observational learning in the rhesus monkey, *J. comp. physiol. Psychol.*, 52, 94–8

Davenport, R. K., Menzel, E. W., and Rodgers, C. M. (1961). Maternal care during infancy. Its effect on weight gain and mortality in the chimpanzee, *Amer. J. Orthopsychiat.*, 31, 803–9

Davis, H., and Danenfeld I. (1967). Extinction induced social interaction in rats, *Psychonomic Science*, 7(3), 85–6

Delgado, J. M. R. (1963). Cerebral heterostimulation in a monkey colony, *Science*, 141 (whole no. 3576), 161–3

Denenberg, V. H., Swain, P. B., Frommer, G. P., and Ross, S. (1958). Genetic physiological and behavioural background of reproduction in the rabbit. IV, An analysis of maternal behaviour at successive parturitions, *Behaviour*, 13, 131–42

Deneberg, V. H., Ottinger, D. R., and Stephens, M. W. (1962). Effects of maternal factors upon growth and behaviour in rats, *Child. Dev.*, 33(1), 65–71

Deneberg, V. H. (1967). Stimulation in infancy, emotional reactivity and exploratory behaviour, in *Biology and Behaviour, Neurophysiology and Emotion*, ed. D. H. Glass, New York

De Vore, I. (1963). Mother-infant relations in free ranging baboons, in *Maternal Behaviour in Mammals*, ed. H. Rheingold, Wiley, New York

De Vore, I. (1963). *Primate Behaviour*, Holt, Rinehart, New York

Dewsbury, D. A. (1967). Effects of alcohol ingestion on the copulatory behaviour of the male rat, *Psychopharmacologica*, 11(3), 276–81

Dimond, S. J. (1965). Restriction of movement and a negative effect during imprinting, *Animal Behav.*, 13, 101–13

Dimond, S. J. (1968). Effects of photic stimulation before hatching on the development of fear in chicks, *J. comp. physiol. Psychol.*, 65, 320–4

Dimond, S. J. (1970). Visual imput and early social behaviour in chicks, in *Social Behaviour in Animals and Man*, ed. J. H. Crook, Academic Press, New York

Douglis, M. B. (1948). Social factors influencing the hierarchies of small flocks of the domestic hen: interactions between resident and part-time members of organized flocks, *Physiol. Zool.*, 21, 147–82

Edwards, D. A. (1968). Mice: fighting by neonatally androgenized females, *Science*, 161, 1027–8

Eibl-Eibesfeldt, I. (1955). Angeborenes und Erworbenes in Nestbauverhalten der Wanderratte, *Naturwissenschaften*, 42, 633–4

Eibl-Eibesfeldt, I. (1958). Das Verhalten der Nagetiere, *Handb. Zool. Berlin.* Bd. 8, Lfg. 12, Teil 10, 1–88

Eibl-Eibesfeldt, I. (1960). Beobachtungen und Versuche an Anemonenfischen (Amphiprion). Observations and experiments on anemone fish of the Maldive and Nicobar Islands. *Z. Tierpsychol.*, 17, 1–10

Emerson, A. E. (1952). The supraorganismic aspects of the society. *Coll. Int. Cent. Rech. Sci. Paris*, 34, 333–53

Feder, H. H., and Whalen, R. E. (1965). Feminine behaviour in neonatally castrasted and estrogen treated male rats, *Science*, 147 (whole no. 3655), 306–7

Folman, Y., and Dróri, D. (1965). Normal and aberrant copulatory behaviour in male rats (*R. Norvegicus*) reared in isolation, *Animal Behav.*, 13(4), 427–9

Forgays, D. G., and Read, J. M. (1962). Crucial periods for free-environmental experience in the rat, *J. comp. physiol. Psychol.*, 55, 816–18

Freedman, L. V., and Rosvold, H. E. (1962). Sexual aggressive and anxious behaviour in the laboratory macaque, *J. Nerv. Ment. Dis.*, 134(1), 18–27

Fuller, J. L. (1953). Cross-sectional and longitudinal studies of adjustive behaviour in the dog, *Ann. N.Y. Acad. Sci.*, 56, 214–24

Fuller, J. L., and DuBuis, E. M. (1962). The behaviour of dogs, in *The Behaviour of Domestic Animals*, ed. E. G. E. Hafez, Ballière, Tindall & Cox, London

Gallup, G. G. (1965). Aggression in rats as a function of frustrative non-reward in a straight alley, *Psychonomic Science*, 3(3), 99–100

Gallup, G. G. (1966). Mirror image reinforcement in monkeys, *Psychonomic Science*, 5, 39

Gartlan, J. S. (1966). Ecology and behaviour of the vervet monkey, Lolui Island, Lake Victoria, Uganda, Ph.D. thesis, University of Bristol

Gerall, A. A. (1966). Hormonal factors influencing masculine behaviour of female guinea pigs, *J. comp. physiol. Psychol.*, 62, 365–9

Gerall, H. D. (1967). Disruption of the male rat's sexual behaviour induced by social isolation, *Animal Behav.*, 15(1), 54–8

Gilbert, R., and Beaton, J. (1967). Imitation and cooperation by hooded rats, *Psychonomic Science*, 8(12), 43

Goldstein, S. R. (1967). Mirror image as a reinforcer in Siamese fighting fish. A repetition with additional controls, *Psychonomic Science*, 7(12), 331

Goltz, F. (1892). Der Hund ohne Groszhirn, *Arch. f. d. ges. Physiol.*, 51, 570–614

Gontarski, H. (1953). Zur Brutbiologie der Honigbiene, *Z. Bienenforsch*, 2, 7–10

Goodall, J. (1965). Chimpanzees of the Gombe Stream Reserve, in *Primate Behaviour*, ed. I. De Vore, Holt, Rinehart, New York

Lawick-Goodall, J. (1968). The behaviour of free living chimpanzees in the Gombe Stream Reserve, *Animal Behav. Monogr.*

Goy, R. W., Bridson, W. E., and Young, W. C. (1964). Period of maximal susceptibility of the prenatal female guinea pig to masculinizing actions of testosterone propionate, *J. comp. physiol. Psychol.*, 57(2), 166–74

Gray, P. H. (1958). Theory and evidence of imprinting in human infants, *J. Psychol.*, 46, 155–66

Green, P. C., and Gordon, M. (1964). Maternal deprivation its influence on visual exploration in infant monkeys, *Science*, 145 (3629), 292–4

Green, P. C. (1965). Influence of early experience and age on expression of affect in monkeys, *J. genet. Psychol.*, 106(1), 157–71

Griffen, D. R. (1958). *Listening in the Dark. The Acoustic Orientation of Bats and Man*, Yale University Press

Griffen, G. A., and Harlow, H. F. (1966). Effects of three months social deprivation on social adjustment and learning on the rhesus monkey, *Child Development*, 37, 533–47

Guhl, A. M., and Allee, W. C. (1944). Some measurable effects of social organization in flocks of hens, *Physiol. Zool.*, 17, 320–47

Guiton, P. (1959). Socialization and imprinting in brown Leghorn chicks, *Animal Behav.*, 7, 26–34

Hafez, E. S. E., Sumption, L. J., and Jackway, J. S. (1962). The behaviour of swine, in *The Behaviour of Domestic Animals*, ed. E. S. E. Hafez, Ballière, Tindall & Cox, London

Hall, K. R. L. (1965). Social organization of the old world monkeys and apes, *Symp. Zool. Soc. Lond.*, 14, 265–90

Hansen, E. W. (1962). The development of maternal and infant behaviour in the rhesus monkey, doctoral diss., University of Wisconsin, University Microfilms, Ann Arbor, Michigan, 63–653

Hansen, E. W., and Mason, W. A. (1962). Socially mediated changes in lever responding of rhesus monkeys, *Psychol. Rep.*, 11(3), 647–54

Harlow, H. F., and Yudin, H. C. (1933). Social behaviour of primates. I, Social facilitation of feeding in the monkey and its relation to attitudes of ascendance and submission, *J. comp. Psychol.*, 16, 171–85

Harlow, H. F. (1960). Primary affectional patterns in primates *Amer. J. Orthopsychiat.*, 30, 676–84

Harlow, H. F. (1961). The development of affectional patterns in infant monkeys, in *Determinants of Infant Behaviour*, ed. B. M. Foss, vol. 1, Methuen, London

Harlow, H. F., and Harlow, M. K. (1962). The effect of rearing conditions on behaviour, *Bull Menninger Clin.*, 26, 213–24

Harlow, H. F., Harlow, M. K., and Hansen, E. W. (1963). The maternal affectional system of rhesus monkeys, in *Maternal Behaviour in Mammals*, ed. H. L. Rheingold, Wiley, New York

Harlow, H. F., and Harlow, M. K. (1965). The affectional systems, in *Behaviour of Non-human Primates*, ed. A. M. Schrier, H. F. Harlow and F. Stollnitz, Academic Press, New York

Harlow, H. F. (1965). Sexual behaviour in the rhesus monkey, in *Sex and behaviour* ed. F. A. Beach, Wiley, New York

Harris, W. V., and Sands, W. A. (1965). The social organization of termite colonies, in *Social Organization of Animal Communities*, Symp. of the Zool. Soc. of London, no. 14

Hart, B. L. (1967). Sexual reflexes and mating behaviour in the male dog, *J. comp. physiol. Psychol.*, 64(3), 388–99

Hayward, S. C. (1957). Modification of sexual behaviour in the male albino rat, *J. comp. physiol. Psychol.*, 50, 70–3

Hebb, D. O. (1949). *Organization of Behaviour*, Wiley, New York

Hediger, H. (1955). *Studies of the Psychology and Behaviour of Captive Animals in Zoos and Circuses*, Butterworth, London

224 References

Heinroth, O. (1911). Beiträge zur Biologie nahmentlich Ethologie und Psychologie der Anatiden, *Verh. 5 int. orn. Kongr. Berlin*, 1910, 589–702

Hersher, L., Richmond, J. B., and Moore, U. A. (1963). Maternal behaviour in sheep and goats, in *Maternal Behaviour in Mammals*, ed. H. L. Rheingold, Wiley, New York

Hess, E. H. (1959). Imprinting, *Science*, 130, 133–41

Hess, W. R. (1928). Stammganglien-Reizversuche, *Ber. ges. Physiol.*, 42, 554

Hess, W. R. (1954). *Diencephalon: Autonomic and Extrapyramidal Functions*, Grune & Stratton, New York

Hinde, R. A., Spencer-Booth, Y., and Bruce, M. (1966). Effects of six days maternal deprivation in rhesus, monkeys, *Nature*, 210, 1021–3

Hinde, R. A., and Spencer-Booth, Y. (1966). The effects of separating rhesus monkeys from their mothers for six days, *J. Child Psychol. and Psychiat.*, 7(3–4), 179–97

Hinde, R. A. and Spencer-Booth, Y. (1967). The behaviour of socially living rhesus monkeys in their first two and a half years, *Animal Behav.*, 15, 169–96

Hoffman, H. S., Stratton, J. W., and Newby, V. (1969). Punishment by response—contingent withdrawal of an imprinted stimulus, *Science*, 163 (3868)

Holder, E. E. (1958). Learning factors in social facilitation and social inhibition in rats, *J. comp. physiol. Psychol.*, 50, 228–32

Horel, J. A., Treichler, R. F., and Meyer, D. R. (1963). Coercive behaviour in the rhesus monkey, *J. comp. physiol. Psychol.*, 56(1), 208–10

Howard, E. (1920), *Territory in Bird Life*, John Murray, London

Hudgens, G. A., Denenberg, V. H., and Zarrow, M. X. (1968). Mice reared with rats: effects of preweaning and postweaning social interactions upon adult behaviour, *Behaviour*, 30(4), 259–74

Hutchinson, R. R., Ulrich, R. E., and Azrin, N. H. (1965). Effects of age and related factors on the pain—aggression reaction, *J. comp. physiol. Psychol.*, 59(3), 365–9

Hutt, C., and Vaizey, M. J. (1966), Differential effects of group density on social behaviour, *Nature*, 209, 1371–2

Hutt, C., and McGrew, L. C. (1968). Some effects upon social behaviour in humans, *A.S.A.B. Symposium on Changes in Behaviour with population Density*, Oxford, July, 1967

Imanishi, K. (1957). Social behaviour in Japanese monkeys, *Macacca fuscata, Psychologia*, 1, 47–54

James, H. (1959). Flicker an unconditioned stimulus for imprinting, *Can. J. Psychol.*, 13, 59–67

James, W. T. (1949). Dominant and submissive behaviour in puppies as indicated by food intake, *J. genet. Psychol.*, 75, 33–43

Jay, P. (1963). Mother-infant relations in langurs, in *Maternal Behaviour in Mammals*, ed. H. L. Rheingold, Wiley, New York

Jensen, G. D., and Tolman, C. W. (1962). Activity level of the mother monkey, *Macacca nemestrina*, as affected by various conditions of sensory access to the infant following separation, *Animal Behav.*, 10(3–4), 228–30

Jensen, G. D. (1965). Mother-infant relationships in the monkey *Macacca nemestrina*. Development of specificity of maternal response to own infant, *J. comp. physiol. Psychol.*, 59(2), 305–8

Kahn, M. W. (1951). The effect of severe defeat at various age levels on the aggressive behaviour of mice, *J. genet. Psychol.*, 79, 117–30

Kahn, M. W. (1954). Infantile experience and mature aggressive behaviour of mice, *J. genet. Psychol.*, 84, 65–75

Kahn, M. W. (1961). The effect of socially learned aggression or submission on the mating behaviour of c.57 mice, *J. genet. Psychol.*, 98, 211–17

Kalsoven, L. G. E. (1936). Onze Rennis van de Javaansche Termieten, *Hand. ned. ind. natuurw. Congr.*, 7, 427–35

Kanek, N. J., and Davenport, D. G. (1967). Between-subject competition: a rat race, *Psychonomic Science*, 7(12), 87

Kanishi, M. (1963). The role of auditory feedback in the vocal behaviour of the domestic fowl, *Z. f. Tierpsychol.*, 20, 349–67

Karli, P. (1958). Hormones stéroides comportement d'aggression interspécifique rat–souris, *J. Physiol. Path. gen.*, 50, 346–7

Karli, P., and Vergnes, M. (1964). Nouvelles données sur les bases neurophysiologiques du comportement d'aggression interspécifique rat–souris, *J. de Physiologie*, 56(3), 384

Kaufman, I. C., and Rosenblum, L. A. (1967). Depression in infant monkeys separated from their mothers, *Science*, 155 (3765), 1030–1

Kaufmann, J. H. (1966). Behaviour of infant rhesus monkeys and their mothers in a free-ranging band, *Zoologica*, 51, 17–28

Kaufman, R. S. (1953). Effects of preventing intromission upon sexual behaviour of rats, *J. comp. physiol. Psychol.*, 46, 209–11

Kellog, W. N., Kohler, R., and Morris, H. N. (1953), Porpoise sounds as sonar signals, *Science*, 117, 239–43

Kelly, A. H., Beaton, L. E., and Magoun, H. W. (1946). A midbrain mechanism for facio-vocal activity, *J. Neurophysiol.*, 9, 181–9

Kennard, M. A. (1955). Effect of bilateral ablation of cingulate area on behaviour of cats, *J. Neurophysiol.*, 18, 159–69

King, J. A. (1956). Sexual behaviour of c.57 BL 10 mice and its relation to early social experience, *J. genet. Psychol.*, 88, 223–9

King, J. A. (1958). Maternal behaviour and behavioural development in two subspecies of *Peromyscus maniculatus*, *J. Mammal.*, 39, 177–90

King, J. A. (1963). Maternal behaviour in Peromyscus, in *Maternal Behaviour in Mammals*, ed. H. Rheingold, Wiley, New York

Klopfer, P. H. (1961). Observational learning in birds: the establishment of behavioural modes, *Behaviour*, 17, 71–80

Krech, D. Rosenzweig, M. R., and Bennett, E. L. (1962). Relations between brain chemistry and problem solving among rats raised in enriched and impoverished environments, *J. comp. physiol. Psychol.*, 55(5), 801–7

Kuehn, R. E., and Beach, F. A. (1963). Quantitative measurement of sexual receptivity in female rats, *Behaviour*, 21(3–4), 282–99

Kuo, Z. Y. (1938). Further study of the behaviour of the cat toward the rat, *J. comp. Psychol.*, 25, 1–8

Kuo, Z. Y. (1967). *The Dynamics of Behavioural Development. An Epigenetic View*, Random House, New York

Lack, D. (1943). *The Life of the Robin*, Penguin, London

Lack, D. (1954). *The Natural Regulation of Animal Numbers*, Oxford University Press

Lanyon, W. E. (1957). The comparative biology of the meadow-larks in *Wisconsin, Nuttall Ornith. Club*, no. 1, Cambridge, Mass.

Larsson, K. (1959). The effect of restraint upon copulatory behaviour in the rat, *Animal Behav.*, 7, 23–5

Larsson, K. (1963). Nonspecific stimulation and sexual behaviour in the male rat, *Behaviour, Leiden*, 20(1–2), 110–14

Latané, B., and Glass, D. C. (1968). Social and nonsocial attraction in rats, *J. Personality and Social Psychology*, 9 (2, 1), 142–6

Lavery, J. J., and Foley, P. J. (1963). Altruism or arousal in the rat, *Science*, 140 (3563), 172–3

Leary, R. W., and Moroney, R. J. (1962). The effect of home cage environment on the social dominance of monkeys, *J. comp. physiol. Psychol.*, 55(2), 256–9

Le Boeuf, B. J. (1967). Heterosexual attraction in dogs, *Psychonomic Science*, 7(9), 313–14

Lehrman, D. S. (1953). Problems raised by instinct theories, *Quart. Rev. Biol.*, 28, 337–65

Lehrman, D. S., and Erickson, C. (1962). Experiment quoted in Interaction between internal and external environments in the regulation of the reproductive cycle of the ring dove, in *Sex and Behaviour*, ed. F. A. Beach, Wiley, New York

Lester, D. (1967). Exploratory behaviour of dominant and submissive rats, *Psychonomic Science*, 9(5), 285–6

Levine, S. (1959). Emotionality and aggressive behaviour in the mouse as a function of infantile experience, *J. genet. Psychol.*, 94, 77–83

Levison, P. K., and Flynn, J. P. (1965). The objects attacked by cats during stimulation of the hypothalamus, *Animal Behav.*, 13(2–3), 217–20

Levy, G. W., and Bevan, W. A. (1958). A failure to find social facilitation of audiogenic seizures in the rat., *Animal Behav.*, 6, 43–4

Leyhausen, P. (1965). The communal organization of solitary mammals, in *The Social Organization of Animal Communities*, Symp. Zool. Soc. London, 14, 249–64

Lilly, J. C., and Miller, A. M. (1961). Sounds emitted by the bottlenose dolphin, *Science*, 133, 1689–93

Lindauer, M. (1952). Ein Beitrag zur Frage der Arbeitsteilung im Bienenstaat, *Z. vergl. Physiol.*, 34, 299–345

Lindzey, G., Monosevitz, M., and Winston, H. (1966). Social dominance in the mouse, *Psychonomic Science*, 5(11), 451–2

Lisk, R. D. (1966). Inhibitory centres in sexual behaviour in the male rat, *Science*, 152 (3722), 669–70

Lissmann, H. W. (1958). On the function and evolution of electric organs in fish, *J. exp. Biol.*, 35, 156–91

Long, T. G., and Smith, H. A. (1965). Communication between dolphins in separate tanks by way of an electric acoustic link, *Science*, 150 (3705), 1839–44

Lorenz, K. (1935). Der Kumpan in der Umwelt des Vogels, *J. f. Orn.*, 83, 137–213, 289–413

Lorenz, K. (1952). *King Solomon's Ring*, Crowell, New York

Lorenz, K. (1963) *On Aggression*, Methuen, London

Marsden, H. M. (1968). Agonistic behaviour of young rhesus monkeys after changes induced in social rank of their mothers, *Animal Behav.*, 16(1), 38–44

Mason, W. A. (1960). The effects of social restriction on the behaviour of rhesus monkeys. I, Free social behaviour, *J. comp. physiol. Psychol.*, 53, 582–9

McDonnell, M., and Flynn, J. P. (1966). Sensory control of hypothalamic attack, *Animal Behav.*, 14, 399–405

McDougall, W. (1908). *An Introduction to Social Psychology*, Methuen, London

McIver, R. M., and Page, C. H. (1949). *Society. An Introductory Analysis*, Holt, Rinehart, New York

Meier, G. W. (1965). Other data on the effects of social isolation during rearing upon adult reproductive behaviour in the rhesus monkey (*Macaca mulatta*), *Animal Behav.* 13(2–3), 228–31

Messmer, E., and Messmer, I. Die Entwicklung der Lautäusserungen und einiger Verhaltensweisen der Amsel (*Turdus merula merula L.*) unter natürlichen Bedingungen und nach Einzelaufzucht in schalldichten Räumen, *Z. f. Tierpsychol.*, 13, 341–441

Michael, R. P., and Keverne, E. B. (1968). Pheromones in the communication of sexual status in primates, *Nature*, 218, 746–9

Miller, N. E., and Carmona, A. (1967). Modification of a visceral response salivation in thirsty dogs, by instrumental training with water reward, *J. comp. physiol. Psychol.*, 63(1), 1–6

Miller, R. E., and Banks, J. H., Jr. (1962). The determination of social dominance in monkeys by a competitive avoidance method, *J. comp. physiol. Psychol.*, 55(1), 137–41

Milum, V. G. (1958). The significance of some honeybee dances, *Proc. 10th. Int. Congr. Ent.*, 4, 1085–8

Mitchell, G. D., Ruppenthal, G. C., Raymond, E. J., and Harlow, H. F. (1966). Long-term effects of multiparous and primiparous monkey mother rearing, *Child Development*, 37(4), 781–91

Mitchell, G. D., Harlow, H. F., Griffin, G. A., and Møller, G. W. (1967). Repeated maternal separation in the monkey, *Psychonomic Science*, 8(5), 197–8

Möhres, F. P. (1957). Elektrische Entadungen in Dienste der Revierabgrenzung, *Naturwissenschaften*, 44, 431–2

Morris, D. (1956). The function and causation of courtship ceremonies in *L'instinct dans le comportement des animaux et de l'homme*, Fondation Singer-Polignac, Paris, 261–86

Morrison, B. J., and Hill, W. F. (1967). Socially facilitated reduction of the fear response in rats raised by groups or in isolation, *J. comp. physiol. Psychol.*, 63 (1), 71–6

Mowrer, O. H. (1940). Animal studies in the genesis of personality, *Trans. N.Y. Acad. Sci.*, 3, 8–11

Mowrer, O. H. (1960). *Learning Theory and the Symbolic Process*, Wiley, New York

Moynihan, M., and Hall, F. (1953). Hostile sexual and other social behaviour patterns of the spice finch (*Lanchura punctulata*) in captivity, *Behaviour*, 7, 33–77

Moynihan, M. (1959). Notes on the behaviour of some North American gulls. IV, The antogeny of hostile behaviour and display patterns, *Behaviour*, 14, 214–39

Moynihan, M. (1962). Hostile and sexual behaviour patterns of South American and Pacific Laridae, *Behaviour*, suppl. 8, 365

Myer, J. S., and Baenninger, R. (1966). Some effects of punishment and stress on mouse killing by rats, *J. comp. physiol. Psychol.*, 62(2), 292–7

Myer, J. S. (1967). Prior killing experience and the suppressive effects of punishment on the killing of mice by rats, *Animal Behav.*, 15, 59–61

Nixon, H. L., and Ribbands, C. R. (1952). Food transmission within the honeybee community, *Proc. Roy. Soc.* (*B*), 140, 43–50

Noirot, E. (1966). Ultrasounds in young rodents. I, Changes with age in albino mice, *Animal Behav.*, 14(4), 459–62

Oldfield-Box, H. (1967). Social organization of rats in a social problem situation, *Nature*, 213 (5075), 533–4

Pearse, A. S. (1914). Habits of fiddler crabs, 1913, *Ann. Rep.*, Smithson. Inst. Wash., 418–28

Peterson, N. (1960). Control of behaviour by presentation of an imprinted stimulus, *Science*, 132, 1395–6

Pierce, J. T., and Nuttall, D. L. (1961). Self-paced sexual behaviour in the female rat, *J. comp. physiol. Psychol.*, 54, 310–13

Pishkin, V., and Shurley, J. T. Social facilitation and sensory deprivation in operant behaviour of rats, *Psychonomic Science*, 6(7), 335–6

Plotnik, R. J., King, F., and Roberts, L. (1965). An objective analysis of social dominance in the squirrel monkey, *Proc. 73rd Ann. Conv. A.P.A.*, 109–10

Pontius, A. A. (1967). A neuro-psychiatric hypothesis about territorial behaviour, *Percept. and Mot. Skills*, 24 (3, 2), 1232–4

Portmann, A. (1964). *Animals as Social Beings*, Harper Torchbooks, New York

Powers, J. B. and Zucker, I. (1967). Estrous behaviour in the pseudo-pregnant and pregnant rat, *Psychonomic Bull.*, 1(2), 22

Presley, W. J., and Riopelle, A. J. (1959). Observational learning of an avoidance response, *J. genet. Psychol.*, 95, 251–4

Pribram, K. H., and Bagshaw, M. (1953). Further analysis of the temporal lobe syndrome utilizing fronto-temporal ablations, *J. comp. Neurol.*, 99, 347–75

Pukowski, E. (1933). Ökologische Untersuchungen an Necrophorus, F., *Z. Morph. Okol. Tiere.*, 27, 518–86

Rabadeau, R. G. (1963). Development of sexual behaviour in the male hamster, *Canad. J. Psychol.*, 17(4), 420–9

Radlow, R., Hale, E. B., and Smith, W. I. (1958). Note on the role of conditioning in the modification of social dominance, *Psychol. Rep.*, 4, 579–81

Reynolds, H. H. (1963). Effect of rearing and habitation in social isolation on performance of an escape task, *J. comp. physiol. Psychol.*, 56(3), 520–5

Ribbands, C. R. (1953). *The Behaviour and Social Life of Honeybees*, Bee Research Association, London

Rice, G. E., and Grainer, P. (1962). 'Altruism' in the albino rat, *J. comp. physiol. Psychol.*, 55(1), 123–5

Rice, G. E. (1965). Aiding responses in rats not in guinea pigs, *Proc. 73rd Ann. Conv. A.P.A.*

Rheingold, H. L. (1963). Maternal behaviour in the dog, in *Maternal Behaviour in mammals*, ed. H. Rheingold, Wiley, New York

Richards, M. P. (1966). Maternal behaviour in virgin female golden hamsters (Mesocricetus auratus waterhouse). The role of the age of the test pup, *Animal Behav.*, 14(2–3), 303–9

Richards, M. P. (1966). Maternal behaviour in the golden hamster: responsiveness to young in virgin, pregnant and lactating females, *Animal Behav.*, 14(2–3) 310–13

234   *References*

Roberts, W. W., and Keiss, H. O. (1964). Motivational properties of hypothalamic aggression in cats, *J. comp. physiol. Psychol.*, 58(2), 187–93

Rosenblatt, J. S. (1953). Mating behaviour in the male cat. The role of social and sexual experience, doct. diss., New York University

Rosenblatt, J. S., and Lehrman, D. S. (1963). Maternal behaviour in the laboratory rat, in *Maternal Behaviour in Mammals*, ed. H. Rheingold, Wiley, New York

Rosenblatt, J. D. (1965). Effects of experience on sexual behaviour in male cats, in *Sex and Behaviour*, ed. F. A. Beach, Wiley, New York

Rosenblatt, J. S. (1965). The basis of synchrony in the behavioural interaction between the mother and her offspring, in the laboratory rat, in *Determinants of Infant Behaviour*, vol. III, ed. B. M. Foss, Methuen, London

Rösch, G. A. (1925). Untersuchungen über die Arbeitsteilung im Bienenstaat. I, Die Tätigkeiten im normalen Bienenstaate und ihre Beziehungen zum Alter der Arbeitsbienen, *Z. vergl. Physiol.*, 2, 571–631

Rosen, J. (1964). Effects of early social experience upon behaviour and growth in the rat, *Child Development*, 35(3), 993–8

Ross, S., Denenberg, V. H., Frommer, G. P., and Swain, P. B. (1959). Genetic physiological and behavioural background of reproduction in the rabbit. Non-retrieving of neonates, *J. Mammal.*, 40, 91–6

Sackett, G. P. (1967). Some effects of social and sensory deprivation during rearing on behavioural development in monkeys, *Revista Interamericana de Psicologia*, 1(1), 55–80

Salazar, J. (1968). Gregariousness in young rats, *Psychonomic Science*, 10(11), 391

Salzen, E. A. (1963). Visual stimuli eliciting the smiling response in the human infant, *J. genet. Psychol.*, 102, 51-4

Schaefer, H. H., and Hess, E. H. (1959). Colour preferences in imprinting, *Z. Tierpsychol.*, 16, 161-72

Schein, M. W. (1963). On the irreversibility of imprinting, *z. Tierpsychol.*, 20, 462-7

Schjelderup-Ebbe, T. (1935). Social behaviour in birds, in *Handbook of Social Psychology*, ed. C. Murchinson, Clark University Press, Worcester, Mass.

Schneirla, T. C. (1952). Basic correlations and coordinations in insect societies with special reference to ants, *Colloq. Int. Cent. Nat. Rech. Sci. Paris*, 34, 247-69

Schneirla, T. C., Rosenblatt, J. S., and Tobach, E. (1963). Maternal behaviour in the cat, in *Maternal Behaviour in Mammals*, ed. H. Rheingold, Wiley, New York

Schutz, F. (1963a). Objectfixierung geschlechtlicher Reaktionen bei Anatiden und Hühnern, *Naturwissenschaften*, 50, 624-5

Schutz, F. (1963b). Über geschlechtlich unterscheidliche Objectfixierung sexueller Reaktionen bei Enten in Zusammenhang mit dem Prachtkleid des Männchens, *Verh. dt. zool. Ges.*, 282-7

Schwartz, M. (1956). Instrumental and consummatory measures of sexual capacity in the male rat, *J. comp. physiol. Psychol.*, 49, 328-33

Scott, J. P., and Frederickson, E. (1951). The causes of fighting in mice and rats, *Psychol. Zool.*, 24, 273-309

Scott, J. P. (1968). *Early Experience and the Organization of Behaviour*, Brooks Cole, California

Seay, B., Hanson, E. W., and Harlow, H. F. (1962). Mother-infant separation in monkeys, *J. Child. Psychol. Psychiat.*, 3, 123–32

Sebeok, T. A. (1965). Animal communication, *Science*, 147 (3661), 1006–14

Sheffield, F. D., Wulff, J. J., and Barker, R. (1951). Reward value of copulation without sex drive reduction, *J. comp. physiol. Psychol.*, 44, 3–8

Shelley, H. P., and Hoyenga, K. T. (1967). Sociability behaviour and the social environment, *Psychonomic Science*, 8(12), 501

Shipley, W. U. (1963). The demonstration in the domestic guinea pig of a process resembling classical imprinting, *Animal Behav.*, 11, 470–4

Simmel, E. C. (1962). Social facilitation of exploratory behaviour in rats, *J. comp. physiol. Psychol.*, 55(5), 831–3

Simpson, J. (1957). Observations on colonies of honey bees subjected to treatments designed to induce swarming, *Proc. R. Ent. Soc. London (A)*, 32, 185–92

Simpson, M. J. A. (1968). The display of the Siamese fighting fish, *Betta Splendens*, *Animal Behav. Monogr.*, vol. 1, part 1

Skinner, B. F. (1938). *The Behaviour of Organisms*, Appleton-Century-Crofts, New York

Skinner, B. F. (1962). Two synthetic social relations, *J. exp. Anal. Behav.* 5(4), 531–3

Sluckin, W. (1964). *Imprinting and Early Learning*, Methuen, London

Smith, F. V., Van-Toller, C., and Boyes, T. (1966). The critical period in the attachment of lambs and ewes, *Animal Behav.*, 14(1), 120–5

Smith, H. (1969). Personal communication, Psychological Laboratory, University College, Cardiff

Sokolova, L. M. (1940). A study of conditional sexual reflexes in rams, *Trud. Lab. inskusst. Oseme. Zivotn.*, (Moscow), 1, 23–35

Soulairac, A. (1952). La signification physiologique de la période refractare dans le compartment sexuel du rat mâle, *J. Physiol. Path. Gen.*, 44, 99–113

Southern, N. H. (1940). The ecology and population dynamics of the wild rabbit, *Oryctolagus cuniculus*, *Ann. appl. Biol.*, 27, 509–26

Southwick, C. H. (1962). Patterns of intergroup social behaviour in primates with special reference to rhesus and howling monkeys, *Ann. N.Y. Acad. Sci.*, 102(2), 436–54

Spitz, R. A., and Wolf, K. M. (1946). The smiling response: a contribution to the ontogenesis of social relations, *Genetic Psychol. Monogr.*, 34, 57–125

Sturman-Hulbe, M., and Stone, C. P. (1929). Maternal behaviour in the albino rat, *J. comp. Psychol.*, 9, 203–37

Swain, P. B., Denenberg, V. H., Ross, S., Hafter, E., and Zarrow, M. X. (1960). Maternal behaviour in the rabbit: hair loosening during gestation, *Amer. J. Physiol.*, 198, 1099–102

Tedeschi, R. E., Tedeschi, D. H., Mucha, H., Cook, L., Mattis, P. A., and Fellows, E. J. (1959). Effects of various centrally active compounds on fighting behaviour in mice, *J. Pharmacol. exp. Ther.*, 125, 28–4

Thach, J. S. (1965). Comparison of social and nonsocial reinforcing stimuli, *Proc. 73rd. A.P.A. meeting*

Thiessen, D. D. (1966). Role of physical injury in the physiological effects of population density in mice, *J. comp. physiol. Psychol.*, 62(2), 322–4

Thompson, T. (1963). Visual reinforcement in Siamese fighting fish, *Science*, 141, 55–7

Thompson, T., and Bloom, W. (1966). Aggressive behaviour and extinction—induced response-rate increases, *Psychonomic Science*, 5(9), 3356

Thompson, W. R. (1958). Social behaviour, in *Behaviour and Evolution* ed. A. Roe and G. G. Simpson, Yale University Press

Thorpe, W. H. (1954). The process of song learning in the chaffinch as studied by means of the sound spectrograph, *Nature*, 173, 465–9

Thorpe, W. H. (1958). The learning of song patterns by birds with special reference to the song of the chaffinch, *Fringilla coelebs*, *Ibis*, 100, 535–70

Thorpe, W. H. (1961). *Bird Song*, Cambridge University Press

Thorpe, W. H. (1965). The ontogeny of behaviour, in *Ideas in Modern Biology*, ed. J. A. Moore, Natural History Press, New York

Tinbergen, N. (1951). *The Study of Instinct*, Clarendon Press, Oxford

Tinbergen, N. (1953). *Social Behaviour in Animals*, Methuen, London

Tinbergen, N. (1957). The functions of territory, *Bird Study*, 4, 14–27

Tolman, C. W. (1961). Social preferences in the albino rat pup, *Psych. Rep.*, 8, 522

Tollman, J., and King, J. A. (1956). The effects of testosterone propionate on aggression in male and female C57 BI 10 mice, *Brit. J. Anim. Behav.*, 4, 147–9

Trowill, J. A. (1967). Instrumental conditioning of the heart rate in the curarized rat *J. comp. physiol. Psychol.*, 63(1), 7–11

Tsai, L. S. (1963). Peace and cooperation among 'natural enemies': educating a rat-killing cat to cooperate with a hooded rat, *Acta Psychologica Taiwonica*, 5, 1–5

Ullrich, W. (1961). Zur Biologie und Soziologie der Colobus affen (Colobus quereza caudatus Thomas 1885), *D. Zool. Garten*, 25, 305–68

Ulrich, R. E. (1961). Reflexive fighting in response to aversive stimulation, *Diss. Abstr.*, 22, 4421

Valenstein, E. S., Riss, W., and Young, W. C. (1955). Experimental and genetic factors in the organization of sexual behaviour in male guinea pigs, *J. comp. physiol. Psychol.*, 48, 397–403

Vaughan, E., and Fisher, A. E. (1962). Male sexual behaviour induced by intracranial electrical stimulation, *Science*, 137 (3533), 758–60

Vernon, W., and Ulrich, R. (1966). Classical conditioning of pain-elicited aggression, *Science*, 152 (3722), 668–9

Von Holst, E., and von Saint Paul, U. V. (1963). On the functional organization of drives, *Animal Behav.*, 11(1), 1–20

Vowles, D. M. (1952). Individual behaviour patterns in ants, *Adv. Sci.*, 10(37), 18–21

Wallis, D. I. (1962). Aggressive behaviour in the ant, *Formica fusca.*, *Animal Behav.*, 10(3–4), 267–74

Wallis, D. I. (1965). Division of labour in ant colonies, in *Social Organization of Animal Communities*, Symp. Zool. Soc. London, 14

Warriner, C. C., Lemon, W. B., and Ray, T. S. (1963). Early experience as a variable in mate selection, *Animal Behav.*, 11, 221–4

Washburn, M. F. (1908). *The Animal Mind*, Macmillan, New York

Washburn, S. L., Jay, P. C., and Lancaster, J. B. (1965). Field studies of old world monkeys and apes, *Science*, 150, 1541–7

Wassman, M., and Flynn, J. P. (1962). Directed attack elicited from hypothalamus, *Arch. Neurol.*, 6, 208–19

Watson, J. B. (1924). *Behaviourism*, University of Chicago Press

Wechkin, S., Masserman, J. H., and Terris, W. (1964). Shock to a conspecific as an aversive stimulus, *Psychonomic Science*, 1(2), 47–8

Weesner, F. M. (1960). Evolution and biology of the termites, *Ann. Rev. Ent.*, 5, 153–70

Welty, J. C. (1934). Experiments in group behaviour of fishes, *Physiol. Zool.*, 7, 85–128

Whalen, R. E. (1963). Sexual behaviour of cats, *Behaviour* 20, 321–4

Whalen, R. E. (1963). The initiation of mating in naive female cats, *Animal Behav.* 11(4), 461–3

Wheeler, L., and Davis, H. (1967). Social disruption of performance on a D.R.L. schedule, *Psychonomic Science*, 7(7), 249–50

Wiesner, B. P., and Sheard, N. M. (1933). *Maternal Behaviour in the Rat*, Oliver & Boyd, London

Williams, E., and Scott, J. P. (1953). The development of social behaviour patterns in the mouse in relation to natural periods, *Behaviour*, 6, 35–64

Willis, F. N. Jr., Michael, G., and Edwards, J. (1966). Persistence of conditioned fighting in a hen pigeon, *Psychonomic Science*, 5(9), 323–4

Wittekindt, E., and Wittekindt, W. (1960). Ein interessante Verhaltensweisse der Sammelbienen, *Leipzig Bienenztg*, 74, 160–3

Wolff, P. C. (1965). The effects of visual impairment on aggressive behaviour, *Psychological Rec.*, 15(2), 185–90

Wynne-Edwards, V. C. (1962). *Animal Dispersion in Relation to Social Behaviour*, Oliver & Boyd, Edinburgh

Zimbardo, P. G. (1958). The effects of early avoidance training and rearing conditions upon the sexual behaviour of the male rat, *J. comp. physiol. Psychol.*, 51, 764–9
Zuckerman, S. (1932). *The Social Life of Monkeys and Apes*, Kegan Paul, London

Author Index

Subject Index